Springer Undergraduate Mathematics Series

Editor-in-Chief
Endre Süli, Oxford, UK

Series Editors
Mark A. J. Chaplain, St. Andrews, UK
Angus Macintyre, Edinburgh, UK
Shahn Majid, London, UK
Nicole Snashall, Leicester, UK
Michael R. Tehranchi, Cambridge, UK

The Springer Undergraduate Mathematics Series (SUMS) is a series designed for undergraduates in mathematics and the sciences worldwide. From core foundational material to final year topics, SUMS books take a fresh and modern approach. Textual explanations are supported by a wealth of examples, problems and fully-worked solutions, with particular attention paid to universal areas of difficulty. These practical and concise texts are designed for a one- or two-semester course but the self-study approach makes them ideal for independent use.

Jolanta Misiewicz

A One-Semester Course on Probability

Jolanta Misiewicz
Department of Mathematics
and Information Science
Warsaw University of Technology
Warsaw, Poland

ISSN 1615-2085　　　　　　ISSN 2197-4144　(electronic)
Springer Undergraduate Mathematics Series
ISBN 978-3-031-86680-7　　　ISBN 978-3-031-86681-4　(eBook)
https://doi.org/10.1007/978-3-031-86681-4

Translation from the Polish language edition: "WYKŁADY Z RACHUNKU PRAWDOPODOBIEŃST WA Z ZADANIAMI" by Jolanta Misiewicz, © JOLANTA K. MISIEWICZ 2013. Published by SCRIPT. All Rights Reserved.

© The Editor(s) (if applicable) and The Author(s), under exclusive license to Springer Nature Switzerland AG 2025

This work is subject to copyright. All rights are solely and exclusively licensed by the Publisher, whether the whole or part of the material is concerned, specifically the rights of reprinting, reuse of illustrations, recitation, broadcasting, reproduction on microfilms or in any other physical way, and transmission or information storage and retrieval, electronic adaptation, computer software, or by similar or dissimilar methodology now known or hereafter developed.
The use of general descriptive names, registered names, trademarks, service marks, etc. in this publication does not imply, even in the absence of a specific statement, that such names are exempt from the relevant protective laws and regulations and therefore free for general use.
The publisher, the authors and the editors are safe to assume that the advice and information in this book are believed to be true and accurate at the date of publication. Neither the publisher nor the authors or the editors give a warranty, expressed or implied, with respect to the material contained herein or for any errors or omissions that may have been made. The publisher remains neutral with regard to jurisdictional claims in published maps and institutional affiliations.

This Springer imprint is published by the registered company Springer Nature Switzerland AG
The registered company address is: Gewerbestrasse 11, 6330 Cham, Switzerland

If disposing of this product, please recycle the paper.

Preface

This book is intended principally for students of mathematics with various specializations, both theoretical and related to applications in finance, insurance, or stochastic modeling. It can be used as the basis for a one-semester lecture course, the participants of which should know mathematical analysis with elements of the theory of Lebesgue measure and integral. However, we do not require from the reader any more advanced knowledge of topology, measure Theory, or functional analysis. The necessary elements of these theories are covered briefly but sufficiently precisely to keep the lecture course complete.

When writing this book, I relied on many well-known and recognized textbooks on the theory of probability and measure theory, in particular the classical *Introduction to Probability Theory and Its Applications* by William Feller [6], and the much more modern and equally extensive book by Jacek Jakubowski and Rafał Sztencel entitled *Introduction to the Probability Theory* [8], which, unfortunately, is available only in Polish. Due to the necessity of limiting the lectures to one semester, I selected only the most important or the most interesting topics. I wanted this textbook not only to provide a good basis for later courses on statistics and stochastic processes but also to be interesting, maybe even amusing at times. Hence, there are several less common problems and exercises, e.g., the Black and White Hats puzzle. I suggest the lecturer should move some extended examples, like Bertrand's Paradox, the Monty Hall Problem, or the Black and White Hats Puzzle, to practice sessions for self-presentation by some students.

Taking into account those students who are not yet familiar with measure theory, I have included those elements of this theory that are necessary to define a probability measure and the expected value as an integral with respect to the measure. Some general, classical results are described in Sects. 8.1 and 8.2. Observing contemporary achievements in the field of stochastic modeling, for example, the evolution of the stock market situation, we find that any simplifications to discrete and absolutely continuous measures are not sufficient. Sections 8.3 and 8.4 contain a systematic definition of conditional expected value based on the Radon–Nikodym Theorem. This is one of the most important theorems in measure theory. We present it here with a partial proof limited to the proof of uniqueness.

Chapter 9 contains hints and answers to some of the exercises. There are no complete solutions since we always want to motivate the reader to act on their own. Solutions to problems which require only simple proofs or calculations are not commented on at all. More interesting exercises interested students can find in [7], [14], [15], and [17]. Those who are more ambitious should take a look at Paul Letac's book [11], where they will find more challenging and demanding exercises.

I would especially like to thank Prof. Czesław Ryll-Nardzewski and Prof. Kazimierz Urbanik, from whom I learned not only the calculus of probability but also the precision and economy of proofs, openness to new problems and that something special that makes mathematics fun. While writing this book, I used the notes from lectures on the theory of real functions delivered by Prof. Cz. Ryll-Nardzewski in the 1970/1971 academic year at the University of Wrocław. I would also like to thank Prof. Anzelm Iwanik, a wonderful mathematician and teacher, to whom I owe my first contact with Probability. I am also grateful to Jacek Bojarski for his assistance in typesetting the manuscript and, with Gosia Mazurek and Karol Bojarski, producing the figures, and to Dorota Stępińska and Oktawia Zegar for proofreading the initial version of the book.

Warsaw, Poland   Jolanta Misiewicz

Declarations

Competing Interests The author has no competing interests to declare that are relevant to the content of this manuscript.

Contents

1 The Beginning 1
    1.1   The Basics of Combinatorics 2
           1.1.1   Permutations 2
           1.1.2   Variations With Repetition 3
           1.1.3   Variations Without Repetition 4
           1.1.4   Combinations 4
    1.2   Putting Objects into Bins 4
           1.2.1   Indistinguishable Balls in Numbered Bins 5
           1.2.2   Distinguishable Balls in Numbered Bins 5
           1.2.3   Indistinguishable Balls in Indistinguishable Bins 6
           1.2.4   Exercises 6

2 The General Definition of Probability 9
    2.1   Families of Sets 9
           2.1.1   Exercises 11
    2.2   The Sample Space 12
           2.2.1   Exercises 12
    2.3   The General Definition of Probability 13
           2.3.1   Basic Properties of Probability 14
           2.3.2   Exercises 17
    2.4   Why the Probability Space $(\Omega, \mathbf{P}, \mathcal{F})$ Must Consist
          of Three Elements 18
           2.4.1   Exercises 20
    2.5   The Classical Definition of Probability 21
           2.5.1   Exercises 22
    2.6   Geometric Probability 25
           2.6.1   Exercises 28
    2.7   Conditional Probability 28
           2.7.1   Exercises 31
    2.8   Independent Events 33
           2.8.1   Exercises 33

	2.9	Bernoulli Trials	34
		2.9.1 Most Probable Number of Successes	35
		2.9.2 The Black and White Hats Puzzle	38
		2.9.3 Exercises	39
	2.10	Upper and Lower Limits of Sequences of Events	40
		2.10.1 Exercises	42
3	**Random Variables and Their Distributions**		45
	3.1	Definition of a Random Variable	45
		3.1.1 Exercises	49
	3.2	Distributions and Cumulative Distribution Functions	50
		3.2.1 Exercises	55
	3.3	Review of Discrete Distributions	56
		3.3.1 Single Point Distributions	56
		3.3.2 Two-Point Distributions	57
		3.3.3 Binomial Distributions or Bernoulli Distributions	57
		3.3.4 Multinomial Distributions	58
		3.3.5 Poisson Distributions	58
		3.3.6 Geometric Distributions	58
		3.3.7 Pascal (Negative Binomial) Distributions	59
		3.3.8 Hypergeometric Distributions	59
		3.3.9 Exercises	60
	3.4	Continuous Type Distributions	60
		3.4.1 Uniform Distributions on an Interval	62
		3.4.2 Exponential Distributions	62
		3.4.3 Gamma Distributions	63
		3.4.4 Beta Distributions	64
		3.4.5 Cauchy Distributions	65
		3.4.6 Gaussian Distributions	65
		3.4.7 Exercises	67
	3.5	A Complete Description of the Types of Random Variables	68
	3.6	Independent Random Variables	71
		3.6.1 Exercises	72
	3.7	Multidimensional Random Variables and Distributions	74
		3.7.1 Exercises	78
4	**Expected Value for Random Variables**		81
	4.1	Expected Value for Simple Random Variables	81
		4.1.1 Exercises	85
	4.2	General Definition of Expected Value	86
		4.2.1 Exercises	95
	4.3	Functions of Random Variables	96
		4.3.1 Exercises	97
	4.4	Expected Value for Continuous Type Random Variables	98
		4.4.1 Exercises	100

	4.5	Expected Value as a Lebesgue–Stieltjes Integral	100
		4.5.1 Exercises	101

5 Random Variable Parameters ... 103
- 5.1 Quantiles, Median, Moments, Variance, Skewness and Kurtosis ... 103
  - 5.1.1 Exercises ... 107
- 5.2 Chebyshev's Inequality ... 108
  - 5.2.1 Exercises ... 110
- 5.3 Parameters of Random Vectors ... 111
  - 5.3.1 Copulas ... 113
  - 5.3.2 H. Markowitz's Investing Theory ... 114
  - 5.3.3 Exercises ... 115
- 5.4 Multivariate Normal Distribution ... 116
  - 5.4.1 Exercises ... 118

6 Characteristic Functions ... 121
- 6.1 Definition and Basic Properties ... 121
  - 6.1.1 Exercises ... 123
- 6.2 Relations Between Distribution, Characteristic Function and Moments of Random Variables ... 125
  - 6.2.1 Exercises ... 130
- 6.3 Weak Convergence of Distributions ... 131
  - 6.3.1 Exercises ... 138
- 6.4 Characteristic Functions of Random Vectors ... 139
  - 6.4.1 Exercises ... 141

7 Limit Theorems ... 143
- 7.1 Kolmogorov's Zero-One Law ... 143
  - 7.1.1 Exercises ... 144
- 7.2 Laws of Large Numbers ... 145
  - 7.2.1 Exercises ... 152
- 7.3 The Central Limit Theorem ... 153
  - 7.3.1 Exercises ... 154

8 Extension of Measure ... 157
- 8.1 The Carathéodory Extension Theorem ... 157
  - 8.1.1 Exercises ... 164
- 8.2 Cumulative Distribution Functions ... 165
- 8.3 The Radon–Nikodym Theorem ... 170
  - 8.3.1 Exercises ... 173
- 8.4 Conditional Expectation ... 173
  - 8.4.1 Conditional Expectation Properties ... 174
  - 8.4.2 Exercises ... 177

9 Hints and Solutions to the Exercises 179

Appendix Table of Normal Distribution Function Φ of $N(0, 1)$ 199

References 201

Index 203

Chapter 1
The Beginning

The birth year of probability theory is usually taken to be 1654. It started like this:

Monsieur de Baussay, alias Antoni Gombauld, best known as Chevalier de Méré, led a lavish life in Paris. From time to time, for pleasure or out of simple curiosity, he visited the gambling salons and instead of falling into a pernicious addiction and gambling fever, he tried to analyze the games on the basis of random or accidental events.

At that time, *the game of six* was especially fashionable. The banker, i.e., a professional gambler hired by the owner of the salon, and the player paid equal stakes to the pool. The winner was the player who failed to roll "6" in four consecutive dice rolls. Chevalier de Méré took a particular interest in the slightly more complicated variation of this game, where two dice were used. He was interested in answering the question: Why is it disadvantageous for the banker to bet that in 24 rolls of two dice, a player will simultaneously roll two sixes?

The number 24 did not appear here by accident. There is the concept of the so-called banker's number, i.e., one where the chances change from favorable for the player to favorable for the banker. It was then believed that since the banker's number when rolling one die was equal to 4, and since rolling two dice gave 6 times more results in one roll, the banker's number when rolling 2 dice was equal to $6 \times 4 = 24$.

On the basis of some theoretical considerations, Chevalier de Méré came to the conclusion that the banker's number for this game was not equal to 24. He got the young French mathematician Blaise Pascal interested in this paradox and it was Pascal who calculated that 24 was still slightly more beneficial to the player, while 25 throws would be slightly more favorable to the banker.

Pascal also solved a more difficult problem, posed again by Chevalier de Méré: the problem of an unfinished game or a partial game. The game consisted of batches and the winner was the player who first won a fixed number of batches. The problem was to determine the fair share of the pot between players when the game was interrupted. Pascal was the first to formulate the principle saying that the winnings of individual participants should depend on the *probability* of each of them winning the game. Based on a few examples he explained precisely how such probabilities should be calculated.

Pascal wrote about his achievements to another French mathematician (and lawyer) Pierre de Fermat. A lively correspondence arose between them on the then-known gambling games, and soon news that Pascal and Fermat had discovered a new branch of mathematics spread around Paris. The literature on probability, or more broadly on probabilistic theory, is vast. A widely known work is W. Feller's book [6] from 1966, which offered the most comprehensive discussion of the foundations of probability at the time. More modern approaches to the subject can be found, for example, in the following books, listed in order of increasing complexity: [3–5, 8, 19]; and for more advanced discussion: [10] and [13].

1.1 The Basics of Combinatorics

In probability theory we often encounter the need to find the number of all possible outcomes of a given experiment, or the number of outcomes that satisfy some additional conditions. This may involve, for example, selection, divisions, or orderings of a finite set of elements. In general, such calculations are not difficult, but to avoid repeating them for each task, we will discuss the most important cases here. Let us begin with the following rule:

Multiplication Principle 1.1 Suppose there are exactly m_1 possible selections of the first element, m_2 possible selections of the second element, ..., and m_k possible selections of the k-th element. If any choice of any element can occur together with any choice of any other element, then the number of all possible choices of k ordered elements is equal to:

$$m_1 \times m_2 \times \cdots \times m_k.$$

Consider, for example, ordering a cake and something to drink in a cafeteria, where they can serve coffee, tea, orange juice, beer, puffs, eclairs, meringues, muffins and cheesecakes. There are 4 ways to choose a drink and 5 ways to choose a cake, thus the number of all possible orders equals $4 \times 5 = 20$.

1.1.1 Permutations

A *permutation* of a set of n distinguishable objects is any ordered arrangement of its elements, numbered with consecutive natural numbers from 1 to n. We can also say that a permutation of n distinguishable elements is a one-to-one function from the set $\{1, \ldots, n\}$ onto the set of elements. If, for example, we have two elements a

and b then their only permutations are:

$$(a,b) \qquad (b,a).$$

All permutations of the elements a, b and c can be obtained by placing the third element c in one of the three possible places in the already defined permutations of a and b; thus we obtain:

$$(c,a,b) \ (c,b,a)$$
$$(a,c,b) \ (b,c,a)$$
$$(a,b,c) \ (b,a,c).$$

Using the same method it is not difficult to prove by mathematical induction that the number of permutations of a set of n distinguishable elements is equal to:

$$1 \cdot 2 \cdot \ldots \cdot n \stackrel{\text{def}}{=} n!$$

Note that we use parentheses to describe a sequence of elements, i.e., an ordered finite set of elements. If the order of elements is irrelevant, we will use braces $\{\ ,\ \}$, e.g. $\{a,b,c\} \equiv \{b,c,a\}$, $\{1,3,7\} \equiv \{1,7,3\}$. We shall use this notation consistently throughout the book.

1.1.2 Variations With Repetition

A *variation with repetition* is any ordered sample of size k from n distinguishable elements where repetition of the same element is allowed. It is therefore any choice of k consecutive elements where the selected element is returned to the set each time—sometimes we say that variations with repetition are return selections. We can also identify a variation with repetition with the corresponding function from the set $\{1, 2, \ldots, k\}$ taking values in the set $\{1, 2, \ldots, n\}$. Words are variations with repetition selected from a set of letters comprising an alphabet. The result of rolling a single die twice is a two-element variation with repetition from the set $\{1, 2, 3, 4, 5, 6\}$. The set of all possible results of rolling the die twice can be described as follows:

$$\{(x, y) : x, y \in \{1, 2, 3, 4, 5, 6\}\}.$$

The easiest way to picture a variation with repetition is to imagine drawing k balls sequentially from a box containing n numbered balls, returning each ball to the box after recording its number. It is clear that there are exactly n ways to choose the first ball, n ways to choose the second ball, etc. Note that the result of the first draw does not affect the result of any other, so by the Multiplication Principle 1.1, we get that the number of all k-element variations with repetitions selected from an n-element set is equal to:

$$W_n^k = n^k.$$

1.1.3 Variations Without Repetition

A *variation without repetition* is defined as any k-element sequence of distinct elements selected from a set n of distinguishable elements, where $n \geq k$. Two variations without repetitions differ from each other either by the elements themselves or by their order in the sequence. Equivalently, we can say that a variation without repetition can be identified with a one-to-one function on the set $\{1, 2, \ldots, k\}$ taking values in the set $\{1, 2, \ldots, n\}$. We have n ways to choose the first element, $n-1$ ways to choose the second, $n-2$ ways to choose the third, and so on. Consequently, the number of all possible k-element variations without repetition taken from an n-element set is equal to:

$$V_n^k = n \cdot (n-1) \cdot \cdots \cdot (n-k+1) = \frac{n!}{(n-k)!}.$$

1.1.4 Combinations

A *k-element combination* of the elements of an n distinguishable element set S is any k-element subset of S. The order of the elements in the fixed combination is irrelevant. For example: the 13 cards which you receive in a bridge deal is a 13-element combination from a 52-element set, the deck of cards.

Combinations differ from variations without repetition only in the fact that in a combination the order in which elements are selected is irrelevant. Note that there are exactly $k!$ k-element variations without repetition containing k fixed elements. Thus, when considering k-element combinations, we shall identify all such variations. This shows that the number of all k-element combinations of an n-element set is equal to:

$$C_n^k = \frac{V_n^k}{k!} = \binom{n}{k}.$$

1.2 Putting Objects into Bins

In statistical physics, one considers the distribution of k particles (balls, elements) into n cells (bins, boxes), where $k \leq n$. Depending on the type of these particles or

1.2 Putting Objects into Bins

on the type of chosen mathematical model, one of the following three assumptions is made: (a) particles are indistinguishable (Bose–Einstein statistics); (b) particles are distinguishable (Maxwell–Boltzmann statistics), or (c) particles are indistinguishable, but there can only be one particle in a bin/cell (Fermi–Dirac statistics). In each of these cases the cells are distinguishable. In probability theory, we shall also consider the case of putting indistinguishable particles into indistinguishable cells.

1.2.1 Indistinguishable Balls in Numbered Bins

We want to find the number of ways to distribute k indistinguishable balls into n numbered bins. The easiest way to solve this problem is to refer to a somewhat childish way of drawing numbered bins: it is enough to draw $n + 1$ vertical lines:

$$\underbrace{| \; | \; | \; \ldots \; | \; |}_{n+1 \text{ lines}}$$

Now, in order to put the balls into the bins, all we have to do is to draw k circles in the same line in such a way that at the beginning and at the end we have vertical lines (so that all balls are in bins). This gives us an $(n + k + 1)$-element sequence consisting of k circles and $(n + 1)$ vertical lines, where the first and last elements in the sequence are lines, e.g.:

$$|\,0\,0\,|\,|\,0\,|\,\ldots\,|\,0\,0\,0\,|\,|.$$

It is easy to see now that the number of such sequences is equal to the number of possible choices of k elements from a set of $n + k - 1$ elements, which we identify with the choices of positions where the circle will be placed:

$$\binom{n+k-1}{k} = \binom{n+k-1}{n-1}.$$

1.2.2 Distinguishable Balls in Numbered Bins

In this case, each ball is assigned the number of the bin in which it will be placed, so the number of possible ways to arrange k balls in n bins is equal to the number of functions from the set $\{1, \ldots, k\}$ which take values in $\{1, \ldots, n\}$. This means that it is equal to the number of k-element variations with repetition from an n-element set, which is equal to n^k.

1.2.3 Indistinguishable Balls in Indistinguishable Bins

To describe the distribution of the balls among the bins we need to write down how many bins are empty, how many contain one ball, how many contain two balls, etc. Let's assume, for example, that we need to put 4 balls in 3 bins. We could put all the balls into one bin, which we notate as $4+0+0$. If we put 2 balls in one bin and the remaining two balls in each of the other two bins, then we would describe this situation using the notation $2+1+1$. We can see now that the possible distributions of 4 indistinguishable balls in 3 indistinguishable bins can be described as follows:

$$4+0+0 = 3+1+0 = 2+2+0 = 2+1+1.$$

If there were 4 bins, we would have:

$$4+0+0+0 = 3+1+0+0 = 2+2+0+0 = 2+1+1+0 = 1+1+1+1.$$

It is evident now that the number of possible distributions of k indistinguishable balls in n indistinguishable bins (notation N_n^k) is equal to the number of ways of writing a natural number k as a sum of n natural numbers (we consider zero to be a natural number). We can also discuss the number of ways of writing a natural number k as a sum of *at most n* natural numbers. We can see that:

$$N_3^4 = 3, \quad N_4^4 = 5, \quad N_1^k = 1, \quad N_2^k = \left\lfloor \frac{k}{2} \right\rfloor, \quad N_n^k = 1 \ \forall n \geq k.$$

General formulas for the number of partitions of a natural number into a sum of natural numbers are not known; instead, asymptotic formulas can be found in some combinatorics textbooks.

1.2.4 Exercises

1. Show that $\binom{n}{k} = \binom{n}{n-k}$, $\binom{n}{0} = 1$, $\binom{n}{1} = n$.
2. Prove that $\binom{n}{k} + \binom{n}{k+1} = \binom{n+1}{k+1}$.
3. Prove that $\sum_{k=0}^{n} \binom{n}{k} = 2^n$, $\sum_{k=0}^{n} (-1)^k \binom{n}{k} = 0$.
4. Prove that $\sum_{k=1}^{n} k \binom{n}{k} = n \, 2^{n-1}$.
5. In bridge, we deal 13-card hands from a 52-card deck. How many possible bridge hands are there?
6. Adam is deal a bridge hand. In how many different ways can Adam get exactly seven spades?
7. How many different results are there when you roll a single die twice?
8. In how many ways can three people be accommodated in two double rooms?

1.2 Putting Objects into Bins

9. How many ways are there to seat n people on n chairs (a) lined up; (b) placed around a round table?
10. In how many ways can seven people be divided into two groups?
11. Ewa cuts a tailor's tape (150 cm) into three parts so that the length of each part is a natural number of centimeters. In how many ways can she do this? How many of these methods will allow her to measure her waist if she needs a length of at least 63 cm length?
12. In how many ways can 13 cards be selected from a deck so that all four aces are among the chosen cards?
13. There are 7 flavors of ice cream in an ice cream shop. How many combinations of flavors can an assistant make if a cone holds no more than 5 scoops of ice cream, assuming that an empty cone does not count as an ice-cream
14. How many four-letter words are there if by a "word" we mean any finite ordered sequence of letters? We consider here words written in Polish, using the 31-letter alphabet, which includes the letters ą, ę, ć, ń, ś, ż, ź but does not contain x, v and q.
15. Six swallows are sitting on five power lines joining two electricity pylons. How many different melodies can be played if we treat the swallows on the wires as sheet music on a stave? Consider two cases: excluding chords (playing several notes at the same time) and including them.
16. We have three paint colors available: red, green and blue. How many three-color flags can we paint in which two adjacent fields are of different colors?
17. Find the number of different four-digit numbers

    (a) divisible by 3,
    (b) divisible by 11,
    (c) in which one of the digits is the sum of the others,
    (d) in which adjacent digits are different.

    Hint: A number is divisible by 11 if the sum of the numbers in the even places minus the sum of the numbers in the odd places is divisible by 11.
18. An exam test consists of 12 sentences. For each of them a student will write **T** if he thinks that the sentence is true or **F** if he thinks that the sentence is false. In how many ways may this test be completed by a student who decides to write the answers randomly?
19. A cashier at a shop has two 1-euro coins, three 50-cent coins and three 10-cent coins in the cash register. How many different coin combinations can he give as change? Suppose that he changes the 50-cent coins into 10-cent coins. How many combinations are there now?
20. A group consists of 15 married couples. In how many ways can a four-person delegation be selected from among them if the delegation may not include any couple?
21. For many years, the following task appearing in high school textbooks puzzled teachers and students: in how many ways can six children be paired? Show that there are three correct solutions to this problem. Is the order in a pair important?

What about the order of pairs? Can you think of a situation/situations in which they may be important?

22. In how many ways can a set comprising k one-cent coins and m five-cent coins be stored in n numbered boxes?
23. In how many ways can k indistinguishable balls be arranged in n numbered drawers if $n \geq k$ and only one ball can be kept in each drawer?
24. In how many ways can k indistinguishable balls be arranged in n numbered drawers if $n \leq k$ and at least one ball should be put in each drawer?
25. Let $\mathbb{N}$ denote the set of all non-zero natural numbers and let $\mathbb{N}_0 = \mathbb{N} \cup \{0\}$. For fixed $n, k \in \mathbb{N}$ find the number of natural solutions of the following equation:

$$k = x_0 + \cdots + x_{n-1}, \qquad x_0, \ldots, x_{n-1} \in \mathbb{N}_0.$$

26. Let $k, n \in \mathbb{N}$ and let N_n^k denote the number of possible distributions of k indistinguishable balls into n indistinguishable bins. Prove that

$$\frac{1}{n!}\binom{n+k-1}{n-1} \leq N_n^k \leq \binom{n+k-1}{n-1}.$$

Chapter 2
The General Definition of Probability

2.1 Families of Sets

Let Ω be any non-empty set and let $\mathcal{A}$ be a non-empty family of subsets of the set Ω. By $\emptyset$ we denote the empty set.

$\mathcal{A}$ is a *ring of sets* if:

(1) $A, B \in \mathcal{A} \implies A \cup B \in \mathcal{A}$;
(2) $A, B \in \mathcal{A} \implies A \setminus B \in \mathcal{A}$.

Note that if $\mathcal{A}$ is a ring of sets then:

(3) $\emptyset \in \mathcal{A}$;
(4) $A_1, \ldots, A_n \in \mathcal{A} \implies \bigcup_{k=1}^{n} A_k \in \mathcal{A}$.

For (3) take any $A \in \mathcal{A}$. By Property (2) we have $\emptyset = A \setminus A \in \mathcal{A}$. Property (4) follows from Property (1) by mathematical induction.

The family $\mathcal{A}$ is a *σ-ring of sets* if it is a ring of sets and the following condition is satisfied:

(5) If $A_1, A_2, \ldots \in \mathcal{A}$ are pairwise disjoint then $\bigcup_{i=1}^{\infty} A_i \in \mathcal{A}$.

Remark 2.1 In Property (5) the assumption that the sets $A_1, A_2, \ldots$ are disjoint can be omitted. If $E_1, E_2, \ldots \in \mathcal{A}$, then

$$\bigcup_{k=1}^{\infty} E_k = \bigcup_{k=1}^{\infty} A_k,$$

where $A_1 = E_1, A_2 = E_2 \setminus E_1, A_3 = E_3 \setminus (E_1 \cup E_2), \ldots$ and the sets $A_1, A_2, \ldots$ are pairwise disjoint.

The family $\mathcal{A} \subset 2^{\Omega}$ is a *field (algebra)* if it is a ring and $\Omega \in \mathcal{A}$.

The family $\mathcal{A} \subset 2^{\Omega}$ is a *σ-field (σ-algebra)* if it is a σ-ring and $\Omega \in \mathcal{A}$.

Note that every algebra $\mathcal{A}$ has the following properties:

(6) $A \in \mathcal{A} \Longrightarrow A' \stackrel{\text{def}}{=} \Omega \setminus A \in \mathcal{A}$;
(7) $A_1, \ldots, A_n \in \mathcal{A} \Longrightarrow \bigcap_{i=1}^{n} A_i \in \mathcal{A}$.

If $A \in \mathcal{A}$ and $\Omega \in \mathcal{A}$, then by Property (2) we have that $A' = \Omega \setminus A \in \mathcal{A}$. In order to prove (7) it suffices to note that

$$\bigcap_{i=1}^{n} A_i = \Omega \setminus \bigcup_{i=1}^{n} (\Omega \setminus A_i)$$

and apply Properties (2) and (4).

Examples 2.2

(a) Let $\Omega = \mathbb{N}$ and let $\mathcal{A}$ be the family of all finite subsets of the set Ω. Then $\mathcal{A}$ is a ring, but it is not a σ-ring. It is an algebra, but not a σ-algebra, since Ω is the union of a countable number of single point sets, but it does not belong to $\mathcal{A}$.
(b) If Ω is any non-empty set, and $\mathcal{A} = \{\Omega, \emptyset\}$, then $\mathcal{A}$ is the smallest σ-field of subsets of the set Ω. Of course, $\mathcal{A}$ is also a ring, a field and a σ-field because every σ-field is a ring, a field and a σ-field.
(c) If $\mathcal{A}$ contains all subsets of the set Ω, i.e.,

$$\mathcal{A} = 2^{\Omega} = \left\{ E : E \subseteq \Omega \right\},$$

then $\mathcal{A}$ is the biggest σ-field of subsets of Ω.
(d) Let $n(A)$ be the number of elements in A. If Ω is an infinite set and

$$\mathcal{A} = \left\{ E \subset \Omega : n(E) < \infty \text{ or } n(\Omega \setminus E) < \infty \right\},$$

then $\mathcal{A}$ is a ring and a field, but is not a σ-ring or a σ-field.

Theorem 2.3 *If for every $i \in I$ the family of sets $\mathcal{A}_i$ is a ring (field, σ-ring, σ-field) of subsets of the fixed set Ω, then $\bigcap_{i \in I} \mathcal{A}_i$ is also a ring (field, σ-ring, σ-field). The cardinality of the index set I is arbitrary.*

Proof If $A, B \in \bigcap_{i \in I} \mathcal{A}_i$, then for every $i \in I$, $A, B \in \mathcal{A}_i$. By the definition of a ring, for every $i \in I$ we have that $A \cup B$, $A \setminus B \in \mathcal{A}_i$, thus $A \cup B$, $A \setminus B \in \bigcap_{i \in I} \mathcal{A}_i$. The other conditions can be proved by analogy. □

Theorem 2.4 *Let $\mathcal{K} \subset 2^{\Omega}$. There exists a smallest ring (field, σ-ring, σ-field) of subsets of the set Ω containing $\mathcal{K}$.*

Proof The proofs of all four statements are very similar, so it is enough to find the smallest σ-field containing $\mathcal{K}$. Let

$$I = \left\{ \mathcal{A} : \mathcal{A} \text{ is a } \sigma\text{-field}, \mathcal{K} \subset \mathcal{A} \right\}.$$

2.1 Families of Sets

Note that $I \neq \emptyset$ because $\mathcal{K} \subset 2^\Omega$ and 2^Ω is a σ-field, thus $2^\Omega \in I$. Let

$$\sigma(\mathcal{K}) \stackrel{\text{def}}{=} \bigcap_{\mathcal{A} \in I} \mathcal{A}.$$

By Theorem 2.3 we know that the set $\sigma(\mathcal{K})$ is a σ-field. If there were a smaller σ-field $\mathcal{G}$ containing $\mathcal{K}$, then, by the definition of the set I, we would have $\mathcal{G} \in I$, so

$$\mathcal{G} \subset \sigma(\mathcal{K}) = \bigcap_{\mathcal{A} \in I} \mathcal{A} \subset \mathcal{G},$$

which implies that $\mathcal{G} = \sigma(\mathcal{K})$. □

Remark 2.5 The σ-field $\sigma(\mathcal{K})$ is called the *σ-field generated by the family $\mathcal{K}$*.

2.1.1 Exercises

27. Does the union of two rings (fields, σ-rings, σ-fields) have to be a ring (field, σ-ring, σ-field)?
28. Prove that $\mathcal{A}$ is a σ-field if and only if the following conditions are satisfied:

    (1) $\Omega \in \mathcal{A}$;
    (2) if $A \in \mathcal{A}$, then $A' = \Omega \setminus A \in \mathcal{A}$; item[(3)] if $A_1, A_2, \cdots \in \mathcal{A}$, then $\bigcup_{n=1}^\infty A_n \in \mathcal{A}$.

30. Prove that the σ-field generated by open rectangles $(a, b) \times (c, d), a < b, c < d$ contains all open sets in the plane $\mathbb{R}^2$.
30. Let $\Omega = [0, 1]$. Determine the rings (fields, σ-rings, σ-fields) of subsets of Ω generated by the following classes of sets:

    (a) $\{[0, 2/3], [1/3, 1]\}$;
    (b) $\{[0, 1/2], [1/2, 1]\}$;
    (c) $\{\emptyset\}$;
    (d) the set of all rational numbers in $[0, 1]$.

31. Let Ω be an uncountable set. Describe the ring (field, σ-ring, σ-field) generated by:

    (a) all one-point subsets;
    (b) all countable subsets;
    (c) all uncountable subsets.

32. Let $\mathcal{A}$ be the family of subsets of the set of natural numbers $\mathbb{N}$ containing all finite subsets and their complements. Is $\mathcal{A}$ a σ-field?

33. Let $\mathcal{F}$ be a σ-field of subsets of Ω and let $B \subset \Omega$. Show that

$$\mathcal{F}_B \stackrel{\text{def}}{=} \{B \cap A : A \in \mathcal{F}\}$$

is a σ-field of subsets of the set B.

2.2 The Sample Space

Let Ω be a set with elements denoted by ω. The set Ω is referred to as the *sample space* or the *space of elementary events* and its elements ω are referred to as *elementary events*. The sample space is a primitive notion in the theory of probability. In specific examples, we will identify it as the set of all possible outcomes of a random experiment.

In the previous sections, we have already seen a few examples of sample spaces, so we will limit ourselves here to discussing only one example.

Example 2.6 We toss a coin until it falls on the same side twice in a row. This can happen on the second toss if the sequence of tosses is (O, O) or (R, R). In this notation O denotes getting a tail and R denotes getting a head. It is possible that we would have to toss the coin three times, then the possible sequences are: (O, R, R), (R, O, O); or four times: (O, R, O, O), (R, O, R, R), etc. The space of elementary events can be identified with the following set:

$$\Omega = \{(O, O), (R, R), (O, R, R), (R, O, O), (O, R, O, O), (R, O, R, R), \ldots\}.$$

2.2.1 Exercises

34. Complete the following equivalences:
    (a) $A \cap B = A \cup B \iff A = \ldots$;
    (b) $[A \cap B = \emptyset \wedge A \cup B = \Omega] \iff B = \ldots$;
    (c) $(A \setminus B) \cup (B \setminus A) = \emptyset \iff A = \ldots$;
    (d) $A \setminus B = A \iff A \cap B = \ldots$

35. Simplify the description of the set E:
    (a) $E = (A \cup B) \cap (A' \cup B)$;
    (b) $E = [(A \cup B) \cap C] \cup [(A \cup C) \cap B]$;
    (c) $E = (A \cup B) \cap (B \cup C) \cap (C \cup A)$;
    (d) $E = (A \cup B) \cap (B \cup C)$, when $A \subset B \subset C$.

36. What conditions must the sets A and B satisfy for the following condition to be true: $(A \cup B) \cap (A' \cup B) \cap (A \cup B') = \emptyset$?
37. Are there sets A, B, and C that satisfy the following conditions: $A \cap B \neq \emptyset$, $A \cap C \neq \emptyset$ and $(A \cap B) \setminus C = \emptyset$?
38. We toss a coin three times. Describe the space of elementary events.
39. You choose three cards from a standard deck. Describe the sample space.

2.3 The General Definition of Probability

Let Ω be a space of elementary events and let $\mathcal{F}$ be a σ-field of its subsets.

Definition 2.7 Let $(\Omega, \mathcal{F})$ be a measurable space. A real-valued function **P** defined on the sets of the σ-field $\mathcal{F}$ is called a *probability* or *probability measure* if **P** satisfies the following axioms:

(1) $\mathbf{P}(A) \geqslant 0$ for every $A \in \mathcal{F}$;
(2) $\mathbf{P}(\Omega) = 1$;
(3) (*countable additivity* or σ-*additivity*) for every sequence of pairwise disjoint sets $A_1, A_2, \cdots \in \mathcal{F}$, we have:

$$\mathbf{P}\left(\bigcup_{k=1}^{\infty} A_k\right) = \sum_{k=1}^{\infty} \mathbf{P}(A_k).$$

The ordered triple $(\Omega, \mathcal{F}, \mathbf{P})$ is called a *probability space*. Every set $A \in \mathcal{F}$ is called a *random event*, or simply an *event*. The probability space $(\Omega, \mathcal{F}, \mathbf{P})$ is *complete* if for every $A \subset \Omega$ the following implication holds:

$$\left(A \subset B \in \mathcal{F}, \quad \mathbf{P}(B) = 0\right) \implies A \in \mathcal{F}.$$

In a complete probability space the σ-field $\mathcal{F}$ of random events contains all subsets of measure zero sets. The empty set $\emptyset$ is called the *impossible event*, and the set Ω is called the *sure event*.

The sum (alternative) of two events A and B is the union of these sets $A \cup B = \{x \in \Omega : x \in A \text{ or } x \in B\}$, while the cross section (conjunction) of events A and B is the intersection of these sets: $A \cap B = \{x \in \Omega : x \in A \text{ and } x \in B\}$. The event opposite to the event A is the event $A' = \Omega \setminus A$.

The simplest probability measure is the *one-point measure* (*Dirac measure*) δ_{ω_0}, which assigns the value 1 to a fixed point $\omega_0 \in \Omega$. To be precise:

$$\delta_{\omega_0}(A) = \begin{cases} 1, & \omega_0 \in A; \\ 0, & \omega_0 \notin A. \end{cases}$$

If there exists a finite or at most countable set $A \in \mathcal{F}$ for which $\mathbf{P}(A) = 1$, then we say that the measure $\mathbf{P}$ is *atomic*. Usually, we additionally assume that the set $A \in \mathcal{F}$ does not have any accumulation points, i.e., points $x \in A$ such that for every $\varepsilon > 0$ the ball $\{y \colon \|x - y\| < \varepsilon\}$ contains some elements of A.

2.3.1 Basic Properties of Probability

Property 1 $\mathbf{P}(\emptyset) = 0$.

Proof Note that $\emptyset = \emptyset \cup \emptyset \cup \cdots$ The empty sets are pairwise disjoint, thus, by axiom (3) we have

$$\mathbf{P}(\emptyset) = \mathbf{P}(\emptyset) + \mathbf{P}(\emptyset) + \cdots.$$

Since $\mathbf{P}(\emptyset) < \infty$, we conclude that $\mathbf{P}(\emptyset) = 0$. □

Property 2 (Finite Additivity) *If* $A \cap B = \emptyset$, $A, B \in \mathcal{F}$, *then* $\mathbf{P}(A \cup B) = \mathbf{P}(A) + \mathbf{P}(B)$. *By mathematical induction this property can be extended to the following: if* $A_1, \ldots, A_n \in \mathcal{F}$ *are pairwise disjoint, then*

$$\mathbf{P}(A_1 \cup \cdots \cup A_n) = \mathbf{P}(A_1) + \cdots + \mathbf{P}(A_n).$$

Proof Since $A \cup B = A \cup B \cup \emptyset \cup \cdots$, we have, by axiom (3) and Property 1,

$$\mathbf{P}(A \cup B \cup \emptyset \cup \emptyset \cdots) = \mathbf{P}(A) + \mathbf{P}(B) + \mathbf{P}(\emptyset) + \cdots = \mathbf{P}(A) + \mathbf{P}(B).$$

□

Property 3 *If* $A \subset B$, $A, B \in \mathcal{F}$, *then* $\mathbf{P}(B \setminus A) = \mathbf{P}(B) - \mathbf{P}(A)$.

Proof The events A and $B \setminus A$ are disjoint, thus by Property 2,

$$\mathbf{P}(B) = \mathbf{P}(A \cup (B \setminus A)) = \mathbf{P}(A) + \mathbf{P}(B \setminus A).$$

□

Property 4 *If* $A \subset B$, $A, B \in \mathcal{F}$, *then* $\mathbf{P}(A) \leq \mathbf{P}(B)$.

Proof It is enough to apply Property 3 and axiom (1). □

Property 5 *For every* $A \in \mathcal{F}$, $\mathbf{P}(A) \leq 1$.

Proof Just note that $A \subset \Omega$, then apply Property 4 and axiom (2). □

2.3 The General Definition of Probability

Property 6 *For every $A \in \mathcal{F}$, we have $\mathbf{P}(A') = 1 - \mathbf{P}(A)$.*

Proof This easily follows from Property 2 and axiom (2). Since the events A and A' are disjoint, we have

$$1 = \mathbf{P}(\Omega) = \mathbf{P}(A \cup A') = \mathbf{P}(A) + \mathbf{P}(A').$$

□

Property 7 (Inclusion-Exclusion Formula) *For every choice of events $A_1, \ldots, A_n \in \mathcal{F}$, $n \in \mathbb{N}$, we have*

$$\mathbf{P}\left(\bigcup_{i=1}^n A_i\right) = \sum_{i=1}^n \mathbf{P}(A_i) - \sum_{(i,j):i<j} \mathbf{P}(A_i \cap A_j)$$
$$+ \sum_{(i,j,k):i<j<k} \mathbf{P}(A_i \cap A_j \cap A_k) - \cdots + (-1)^{n-1}\mathbf{P}(A_1 \cap \cdots \cap A_n).$$

Proof Note that the event $A \cup B$ can be written as a sum of two disjoint events: A and $B \setminus A$. By Property 2, we get

$$\mathbf{P}(A \cup B) = \mathbf{P}(A \cup (B \setminus A)) = \mathbf{P}(A) + \mathbf{P}(B \setminus (A \cap B)).$$

Since $(A \cap B) \subset B$, the result for $n = 2$ follows from Property 3. The inclusion-exclusion formula for an arbitrary $n \in \mathbb{N}$ is obtained using mathematical induction.

□

Property 8 (Probability Continuity Theorem) *If $A_n \in \mathcal{F}, n \in \mathbb{N}$, is an increasing sequence of random events, i.e., $A_1 \subset A_2 \subset \cdots$, then*

$$\mathbf{P}\left(\bigcup_{k=1}^\infty A_k\right) = \lim_{n\to\infty} \mathbf{P}(A_n).$$

If $A_n \in \mathcal{F}, n \in \mathbb{N}$, is a decreasing sequence of random events, i.e., $A_1 \supset A_2 \supset \cdots$, then

$$\mathbf{P}\left(\bigcap_{k=1}^\infty A_k\right) = \lim_{n\to\infty} \mathbf{P}(A_n).$$

Notation If (A_n) is an increasing sequence of random events and $A = \bigcup_{n=1}^\infty A_n$, then we write $A_n \uparrow A$. If (A_n) is a decreasing sequence of random events and $A = \bigcap_{n=1}^\infty A_n$, then we write $A_n \downarrow A$.

Proof Let $A_n \in \mathcal{F}$, $n \in \mathbb{N}$, be an increasing sequence of random events. Since the events $A_{n+1} \setminus A_n$, $n \in \mathbb{N}$, are pairwise disjoint, by σ-additivity of probability measure we have

$$\mathbf{P}\left(\bigcup_{k=1}^{\infty} A_k\right) = \mathbf{P}(A_1) + \mathbf{P}(A_2 \setminus A_1) + \mathbf{P}(A_3 \setminus A_2) + \cdots$$

Since the obtained series is convergent, $\sum_{k=n+1}^{\infty} \mathbf{P}(A_k \setminus A_{k-1})$ tends to zero as $n \to \infty$ and we have

$$\mathbf{P}\left(\bigcup_{k=1}^{\infty} A_k\right) = \mathbf{P}(A_1) + \mathbf{P}(A_2) - \mathbf{P}(A_1) + \cdots + \mathbf{P}(A_n) - \mathbf{P}(A_{n-1})$$

$$+ \sum_{k=n+1}^{\infty} \mathbf{P}(A_k \setminus A_{k-1}) = \mathbf{P}(A_n) + \sum_{k=n+1}^{\infty} \mathbf{P}(A_k \setminus A_{k-1})$$

$$= \lim_{n \to \infty} \mathbf{P}(A_n).$$

Assume now that the sequence of random events $A_n \in \mathcal{F}$, $n \in \mathbb{N}$, is decreasing. We see that the sequence $(A_1 \setminus A_n) \in \mathcal{F}$, $n \in \mathbb{N}$, is increasing, thus, using the first statement of this property, we have

$$\mathbf{P}(A_1) - \mathbf{P}\left(\bigcap_{k=1}^{\infty} A_k\right) = \mathbf{P}\left(A_1 \setminus \bigcap_{k=1}^{\infty} A_k\right) = \mathbf{P}\left(\bigcup_{k=1}^{\infty}(A_1 \setminus A_k)\right)$$

$$= \lim_{k \to \infty} \mathbf{P}(A_1 \setminus A_k) = \mathbf{P}(A_1) - \lim_{k \to \infty} \mathbf{P}(A_k).$$

□

Property 9 (Subadditivity of Probability Measure) *For every sequence of probability events* $A_1, A_2, \ldots \in \mathcal{F}$, *the following inequality holds*

$$\mathbf{P}\left(\bigcup_n A_n\right) \leq \sum_n \mathbf{P}(A_n).$$

Proof Note that $B_n = A_n \setminus (A_1 \cup \cdots \cup A_{n-1})$, $n \in \mathbb{N}$, is a sequence of pairwise disjoint events and

$$\bigcup_{n=1}^{\infty} A_n = \bigcup_{n=1}^{\infty} B_n.$$

2.3 The General Definition of Probability

Since $B_n \subset A_n$ for each $n \in \mathbb{N}$, $\mathbf{P}(B_n) \leqslant \mathbf{P}(A_n)$ by Property 4. Now it is enough to apply axiom (3) of probability measure to obtain

$$\mathbf{P}\left(\bigcup_n A_n\right) = \mathbf{P}\left(\bigcup_n B_n\right) = \sum_{n=1}^{\infty} \mathbf{P}(B_n) \leqslant \sum_{n=1}^{\infty} \mathbf{P}(A_n),$$

which was to be shown. □

2.3.2 Exercises

40. A sample space Ω contains exactly n elementary events. What is the smallest and largest possible number of random events in this space?
41. Can the set of random events of a certain probabilistic space Ω consist of (a) 128; (b) 129; (c) 130 elements?
42. Could the number of elementary events in a space Ω be is greater than the number of random events in Ω?
43. Prove the inclusion-exclusion formula for an arbitrary $n \in \mathbb{N}$, assuming that it holds for $n = 2$.
44. Let $\mathcal{F}$ be a σ-field of subsets of the set Ω. Prove that if $\mathbf{P}_1, \ldots, \mathbf{P}_n$ are probability measures on the space $(\Omega, \mathcal{F})$ and $c_1, \ldots, c_n$ are positive constants such that $c_1 + \cdots + c_n = 1$, then $c_1 \mathbf{P}_1 + \cdots + c_n \mathbf{P}_n$ is also a probability measure.
45. Let $x \in \Omega$. Show that the Dirac-delta: $\delta_x(A) \stackrel{\text{def}}{=} \mathbf{1}_x(A)$, i.e., $\delta_x(A) = 1$ if $x \in A$ and $\delta_x(A) = 0$ if $x \notin A$, is a probability measure on $(\Omega, \mathcal{F})$ for any σ-field $\mathcal{F}$ of subsets of Ω.
46. Prove that for a decreasing sequence of events $A_1, A_2, \cdots \in \mathcal{F}$, the following implication holds:

$$\bigcap_{n=1}^{\infty} A_n = \emptyset \implies \lim_{n \to \infty} \mathbf{P}(A_n) = 0.$$

47. Show that if $\mathbf{P}(A) = 0.7$ and $\mathbf{P}(B) = 0.8$, then $\mathbf{P}(A \cap B) \geq 0.5$.
48. Is it true that in any probability space $(\Omega, \mathcal{F}, \mathbf{P})$ and for any $A \in \mathcal{F}$ the following equivalence holds:

$$\mathbf{P}(A) = 0 \iff A = \emptyset \ ?$$

49. Let $(\Omega, \mathcal{F}, \mathbf{P})$ be a probability space and

$$\mathcal{F}_c = \{E \subset \Omega : \ \exists A, B \in \mathcal{F} \ \ (E \setminus A) \cup (A \setminus E) \subset B, \ \ \mathbf{P}(B) = 0\},$$

where the set $E \bigtriangleup A = (E \setminus A) \cup (A \setminus E) = (A \cup E) \setminus (A \cap B)$ is called the *symmetric difference* of the sets A and E. If $E \in \mathcal{F}_c$ and $E \bigtriangleup A$ is a subset of a **P**-zero set, then we define $\mathbf{P}_c(E) = \mathbf{P}(A)$. Show that $(\Omega, \mathcal{F}_c, \mathbf{P}_c)$ is a complete probability space.

50. Let $(\Omega, \mathcal{F}, \mathbf{P})$ be a complete probability space and let $B \notin \mathcal{F}$ be a subset of Ω for which the following implication holds:

$$\left(A \in \mathcal{F}, \ B \subset A \right) \implies \mathbf{P}(A) = 1.$$

We define $\mathcal{F}_B = \{A \cap B : A \in \mathcal{F}\}$ and $\mathbf{P}_B(E) = \mathbf{P}(A)$ when $E = A \cap B \in \mathcal{F}_B$. Show that $(B, \mathcal{F}_B, \mathbf{P}_B)$ is a complete probability space.

2.4 Why the Probability Space $(\Omega, \mathbf{P}, \mathcal{F})$ Must Consist of Three Elements

It might seem that we are overly theorizing here. Would it not be enough to determine the probability $\mathbf{P}$ on all subsets of the set Ω? No extra σ-field is needed! It turns out, however, that such a construction is possible only when the space Ω contains at most countably many elements. And even then, it does not always happen.

To put it simply, the σ-field $\mathcal{F}$ is the class of those subsets of the set Ω whose probability we are able to measure. Two cases may occur here: either the probability of the event A cannot be determined because the probability measure $\mathbf{P}$ available to us is too poor, or the class of sets 2^Ω is too rich. We will discuss this with examples.

Example 2.8 Imagine that we have three apples (elementary events) A, B and C with weights respectively $0.2, 0.3$, and 0.5 kg. We also have an ordinary pan balance and one 0.5 kg weight. In this situation we can conclude that two apples A and B weigh the same as apple C, but we are not able to give their weights. Hence, the class of those subsets for which we can measure their weight is equal to:

$$\{\emptyset, A \cup B, C, A \cup B \cup C\}.$$

Our measuring instrument, and thus the measure it generates, is too poor in this case as it cannot measure all subsets of the set of three apples.

Example 2.9 Let $\Omega = [-1, 2]$ and let λ be the normalized Lebesgue measure on Ω. By Q we denote the set of all rational numbers in the interval $[-1, 1]$ ordered into a sequence:

$$Q = \{q_0 = 0, q_1, q_2, \ldots\} \quad \text{such that } q_i \neq q_j \text{ for } i \neq j.$$

2.4 Why the Probability Space $(\Omega, \mathbf{P}, \mathcal{F})$ Must Consist of Three Elements

Suppose we can construct $E \subset [0, 1]$ such that

$$(E + q_n) \cap (E + q_k) = \emptyset \text{ if } n \neq k, \quad [0, 1] \subset \bigcup_{n=1}^{\infty} (E + q_n) \subset [-1, 2].$$

If the set E were measurable with respect to the measure λ, i.e., if the number $\lambda(E)$ were well defined, then from the invariance of Lebesgue measure under shifts we would have $\lambda(E + q_n) = \lambda(E)$ for every $n \in \mathbb{N}$. Since the sets $(E + q_n)$ are disjoint,

$$\lambda\left(\bigcup_{n=1}^{\infty} (E + q_n)\right) = \sum_{n=1}^{\infty} \lambda(E + q_n) \leq \lambda([-1, 2]) = 1.$$

Hence, we conclude that $\lambda(E) = 0$. Otherwise, the considered series would diverge to infinity. On the other hand, however, we have:

$$\frac{1}{3} = \lambda([0, 1]) \leq \lambda\left(\bigcup_{n=1}^{\infty} (E + q_n)\right) = \sum_{n=1}^{\infty} \lambda(E + q_n) = 0,$$

which is a contradiction, hence such a set E cannot belong to the σ-field of λ-measurable sets.

Construction of the Set E First, we divide $[0, 1]$ into uncountably many disjoint sets. For every $a \in [0, 1]$, we define

$$[a] = \left\{x \in [0, 1] : x - a \in Q\right\}.$$

Of course, $a \in [a]$, so the sets $[a]$ are not empty. Note that for $a, b \in [0, 1]$, we have:

(1) if $a - b \in Q$, then $[a] = [b]$;
(2) if $a - b \notin Q$, then $[a] \cap [b] = \emptyset$.

Indeed, if $a - b \in Q$ and $x \in [a]$, then $x - a \in Q$, so $x - b = (x - a) + (a - b) \in Q$ and $x \in [b]$. Hence, we get that $[a] \subset [b]$, and, due to the symmetry of the assumption, also $[b] \subset [a]$. Assume now that $a - b \notin Q$ and $[a] \cap [b] \neq \emptyset$. Then there exists a number $c \in [0, 1]$ such that $c - a, c - b \in Q$. This, however, implies that $a - b = (c - b) - (c - a) \in Q$. The obtained contradiction implies (2).

Now let $\mathcal{E}$ be the family of all sets $[a]$, $a \in [0, 1]$. It follows from properties (1) and (2) that any two different sets from the family $\mathcal{E}$ are disjoint. Using the axiom of choice, we now construct the set E by including in it one element from each of the sets of the family $\mathcal{E}$, i.e., so that for any $a \in [0, 1]$ the set $[a] \cap E$ is a single point.

It remains to check that the sets $(E + q_n)$ are disjoint for different n. Suppose this is not the case, thus for some $n \neq k$, there exists $c \in (E + q_k) \cap (E + q_n)$. But then $c - q_k, c - q_n \in E$, and since these numbers are different, the construction of

E shows that $[c-q_k] \cap [c-q_n] = \emptyset$. Property (2) implies that $(c-q_k) - (c-q_n) = q_n - q_k \notin Q$. The obtained contradiction concludes the proof of the correctness of the construction.

Remarks 2.10

(1) If we replace the set E by the *Vitali set* $M = \{e^{2\pi ri} : r \in E\}$ (a paradoxical set in S_2) and replace the equivalence classes $[a]$ by $\langle a \rangle = \{e^{2\pi ri} : r \in [a]\}$, then the same construction can be written as a very impressive Vitali theorem: There exists a partition of the unit circle $S_2 \subset \mathbb{R}^2$ into countably many pieces, such that from any infinite subset of pieces we can build the original circle using only rotations. This partition is called a *paradoxical circle partition*. For details, see Vitali [18].
(2) Mazurkiewicz and Sierpiński gave an example of a paradoxical (due to isometry) partition of the plane. For details, see Mazurkiewicz and Sierpiński [12].
(3) The Banach–Tarski theorem concerns the existence of a paradoxical partition of the ball S_3. It states that a three-dimensional sphere can be "cut" into a finite number of parts (five are enough), and then, using only shifts and rotations, two spheres with the same radii as the radius of the initial sphere can be built from these parts. It is paradoxical that on the one hand, as a result of cutting, shifting, rotating and folding operations, the volume of the sphere can be doubled, while on the other hand, the shifting and rotation operations used are isometries and preserve the volume of solids. For details, see Banach and Tarski [2].
(4) Banach and Kuratowski showed even more. They proved, under the continuum hypothesis, that there is no countably additive measure defined on all subsets of $\mathbb{R}$ such that the measure of every single-element set is equal to zero. For details, see Banach and Kuratowski [1].

2.4.1 Exercises

51. What is the smallest and greatest σ-field $\mathcal{F}$ for which $(\mathbb{R}, \mathcal{F}, \delta_1)$ is a probability space?
52. Let $\varepsilon > 0$ and let $f : [-1, 2] \to (\varepsilon, \infty)$ be an integrable function satisfying the condition $\int_{-1}^{2} f(x)\,dx = 1$. We define

$$\mathbf{P}(A) = \int_A f(x)\,dx,$$

for every Borel set $A \subset [-1, 2]$. Prove that the set E constructed in this section is non-measurable with respect to $\mathbf{P}$.
53. Show that every countable set $A \subset [-1, 2]$ is measurable with respect to the measure $\mathbf{P}$ defined in the previous exercise.

2.5 The Classical Definition of Probability

Let Ω be a sample space, that is, the set of all possible mutually exclusive results of a random experiment. In this section we assume that Ω is a finite set. Every subset of the set Ω is a random event, so $\mathcal{F} = 2^\Omega$.

When considering dice rolling, the set Ω can be identified with the six-element set $\{1, 2, 3, 4, 5, 6\}$. In the case of tossing a coin, it is a two-element set $\Omega = \{head, tail\}$ or, if we identify *head* with 1 and *tail* with 0, $\Omega = \{0, 1\}$. If we are considering rolling a single die twice, then the elementary event can be described as a pair (a, b), where a is the number obtained on the first roll and b on the second; then

$$\Omega = \{(a, b) \colon a, b \in \{1, 2, 3, 4, 5, 6\}\}.$$

If, on the other hand, we consider the simultaneous selection of two balls from an urn containing six numbered balls, then

$$\Omega = \{\{a, b\} \colon a, b \in \{1, 2, 3, 4, 5, 6\}, a \neq b\}.$$

We have presented here a few examples of the formal description of a sample space. This description is very important in probability theory and in calculating probabilities of random events—it is crucial to consequently describe a random event as a subset of the given set Ω.

Definition 2.11 Suppose Ω consists of n elements of equal importance (equally probable). The probability of any event $A \in 2^\Omega$ is given by the formula:

$$\mathbf{P}(A) = \frac{\text{number of elements in } A}{n} = \frac{n(A)}{n(\Omega)}.$$

This is the so-called *classical definition of probability*. Let us emphasize once again that it only applies if Ω contains finitely many equally probable elementary events. As we will see later, this is an exceptional case.

Example 2.12 If, in a two-dice roll, the event A consists of the results whose sum is divisible by 5, then

$$\Omega = \{(a, b) \colon a, b \in \{1, 2, 3, 4, 5, 6\}\}, \qquad n(\Omega) = 36,$$
$$A = \{(1, 4), (4, 1), (2, 3), (3, 2), (5, 5), (4, 6), (6, 4)\}, \quad n(A) = 7.$$

Since there is no reason to suspect that any of the elementary events will happen more often than any other, we use the classical definition of probability and obtain $\mathbf{P}(A) = 7/36$.

Similarly, we can define $\Omega = \{2, 3, 4, \ldots, 12\}$ as the sample space because the sum of the obtained results must be a number from this set. Note, however, that now the elementary events from Ω are not equally probable. For example, the sum 2 can be obtained only when we get 1 on both dice, while there are three ways to obtain a sum of 4: $1+3 = 2+2 = 3+1$. This implies that the sum 4 is three times more probable than the sum 2, so the elementary events are not equally probable in the sample space defined in this way.

It is easy to see that the classical probability **P** as a function on the set $\mathcal{F} = 2^\Omega$ has the following properties:

(i) $\forall A \in \mathcal{F}$ $\qquad\qquad 0 \leqslant \mathbf{P}(A) \leqslant 1$;
(ii) $\mathbf{P}(\emptyset) = 0,$ $\qquad\qquad \mathbf{P}(\Omega) = 1$;
(iii) $\forall A, B \in \mathcal{F}, A \cap B = \emptyset \qquad \mathbf{P}(A \cup B) = \mathbf{P}(A) + \mathbf{P}(B)$;
(iv) $\forall A \in \mathcal{F}$ $\qquad\qquad \mathbf{P}(A) = 1 - \mathbf{P}(A')$.

2.5.1 Exercises

54. We have eight casino chips in a box: $2, 4, 6, 7, 8, 11, 12, 13$. What is the probability that the fraction obtained by dividing the numbers of two randomly taken chips is irreducible.

55. **The Genoese Lottery.** A lottery ticket has a table with numbers from 1 to 90. By paying a fixed stake for each selected number, a player can mark k numbers from the table, $k = 1, 2, 3, 4, 5$. Then, by drawing lots, five numbers are selected from 1 to 90. If it turns out that all the numbers crossed out by the player are among the five drawn, his win is as follows:

$k = 1$ $\qquad$ 15 stakes;
$k = 2$ $\qquad$ 270 stakes;
$k = 3$ $\qquad$ 5500 stakes;
$k = 4$ $\qquad$ 75000 stakes;
$k = 5$ $\qquad$ 1000000 stakes.

Calculate the probability of winning the lottery in each of these cases.

56. We draw 5 domino-stones out of the set of 28. What is the probability that among the drawn stones there is at least one such that the sum of its two fields (i.e. the total number of pips on the stone) is equal to 6?

57. We throw n dice. What is the probability of getting the same result on each of them?

58. There are $n + k$ seats in a cinema room and exactly n viewers have come to the screening. If they choose their seat randomly, what is the probability that all m places, $m < n$, in the fifth row will be occupied?

59. To limit the number of semi-final matches, $2n$ soccer teams are split into two equal groups. What is the probability that the two strongest teams will end up in the same group?

2.5 The Classical Definition of Probability 23

60. From an urn containing $2n$ white balls and $2n$ black balls, we randomly draw $2n$ balls, returning each ball to the urn after it has been drawn. What is the probability that the same number of white and black balls are drawn? What would this probability be if the draw was done without returning the balls back to the urn?
61. We have five balls in an urn: colored white, black, red, green and white. We draw a single ball 25 times, each time returning it into the urn. What is the probability that each color appears exactly 5 times?
62. We draw without replacement a few balls from an urn which contains white and black balls. What is more probable: drawing a white ball first or drawing a white ball last?
63. There are n black balls and αn white balls in an urn. We draw all the balls one by one from the urn. What is the probability that the last drawn ball is black?
64. There are black and white balls in an urn. Prove that the probability of drawing (with replacement) two balls of the same color is not less than 0.5. Is this claim true when the draw is done without replacement?
65. There are n balls in N numbered boxes, $n < N$. What is the probability that in each box there is at most one ball if the balls are (a) indistinguishable, (b) distinguishable.
66. We arrange 30 numbered balls in 8 numbered drawers.

    (a) Calculate the probability that 3 drawers will remain empty, 2 drawers contain 3 balls, 2 drawers contain 6 balls and the remaining 12 balls are in one drawer.
    (b) What would the probability be if the balls were indistinguishable?

67. A postman is to deliver 30 letters to eight apartments numbered 1 to 8. If all addresses are equally likely to appear on the envelopes, what is the chance that in apartment 5 the postman will leave exactly k letters, $k = 1, \ldots, 30$?
68. We randomly throw n coins into $n - 1$ money boxes. Calculate the probability that the pink piggy-bank will be empty.
69. $n + 2$ items are randomly placed into n boxes. What is the probability that at most one box is empty?
70. $N+1$ items are randomly placed into n numbered boxes. What is the probability that exactly two boxes are empty?
71. $2n$ items are put into n boxes. Calculate the probability that none of the boxes is empty.
72. We have $2n$ sheets of paper and $2n$ envelopes. The sheets and envelopes are separately numbered from 1 to $2n$. We randomly put each sheet of paper into an envelope, so that each envelope contains one sheet. What is the probability that the sum of the numbers on the envelope and on the sheet put into it will be odd?

73. We have a well-shuffled deck of 52 cards. Calculate the probability that:

    (a) the first four cards in the deck are aces;
    (b) the first and last cards in the deck are aces;
    (c) aces in the deck are separated by exactly k other cards.

74. Tom, his partner and another couple are playing bridge. They are each dealt a 13-card hand. What is the chance that

    (a) each player will get one ace?
    (b) one of the players will get a full suit?
    (c) each player will be dealt all cards from deuce to ace, possibly of different suits?
    (d) Tom will receive exactly n spades, and his partner exactly m spades?

75. Eight chess rooks are placed on a chessboard on randomly selected fields. What is the probability that none of them can capture the other and none of them stands on the main diagonal of white squares?

76. From a deck of 52 cards 13 have been randomly selected. What is the probability that exactly k pairs (*ace, king*) of the same suit are among the chosen ones?

77. Adam is dealt five cards in a poker game. What is the chance that Adam will get a straight flush?

78. From a batch of N goods, including M which comply with the standard, we draw n items: (a) with replacement; (b) without replacement. Calculate the probability that among the randomly selected goods there will be exactly k which are compliant with the standard.

79. Seven bridges connect Burghers' Island in Wrocław with the town. What is the probability that two friends will meet if one of them is just entering the island and the other is leaving it?

80. In a dark room, we have a basket containing n pairs of shoes. We choose k shoes and move them to a well-lit corridor. What is the probability that among the chosen shoes there are exactly r complete pairs, $2r \leqslant k \leqslant 2n$?

81. A solar system consists of one sun, four planets, and five moons. What is the probability that no moon revolves around one of these planets?

82. Each of five electrons orbits one of four atomic nuclei. What is the probability that no electron revolves around one of these nuclei?

83. Six tourists stay overnight in a mountain shelter which has three guest rooms: a double, triple and quadruple room. What is the probability that one of the rooms has been left vacant? And what answer to this question would be given by the manager of the shelter who has not seen either the tourists or their accommodation?

2.6 Geometric Probability

Recall that the *σ-field of Borel sets*, or *Borel σ-field* $\mathcal{B}(\mathbb{R}^k)$ in the space $\mathbb{R}^k$, is the σ-field generated by the family of open balls

$$K(x_0, r) = \{x \in \mathbb{R}^k : d(x, x_0) < r\}, \quad x_0 \in \mathbb{R}^k, r > 0,$$

where d denotes the Euclidean distance in $\mathbb{R}^k$. For simplicity, we write $\mathcal{B}(\mathbb{R}) = \mathcal{B}$. Let λ_k denote the Lebesgue measure proper for $\mathbb{R}^k$, i.e., λ_1 measures the length of intervals, λ_2 measures the area of plane figures, and λ_3 measures the volume of solids in $\mathbb{R}^3$.

Assume that the sample space Ω is a Borel subset of $\mathbb{R}^k$ such that $0 < \lambda_k(\Omega) < \infty$. We assume that the σ-field of random events is equal to the set of all Borel subsets in Ω, i.e., $\mathcal{F} = \mathcal{B}(\Omega)$. The probability of any random event $A \in \mathcal{F}$ is then defined by:

$$\mathbf{P}(A) = \frac{\lambda_k(A)}{\lambda_k(\Omega)}.$$

This is called the *geometric probability*.

Example 2.13 (Buffon's Needle) The problem of Buffon's needle is one of the most interesting applications of geometric probability. Let's randomly throw a needle of length $2r$, $2r < l$, onto a plane lined with parallel lines at distance l apart. In order to calculate the probability that the needle will touch one of the straight lines, we must first unambiguously describe the position of the needle in relation to the nearest straight line (see Fig. 2.1).

By x we denote here the distance from the center of the needle to the nearest straight line below. Let φ be the measure of the angle the needle makes with this line. It is easy to see that x may take any value in the interval $[0, l)$, and φ can take values in $[0, \pi)$. The sample space and its measure are therefore given by

$$\Omega = [0, \pi) \times [0, l), \qquad \lambda_2(\Omega) = \pi l.$$

Fig. 2.1 Buffon's needle: the position of the needle in relation to the nearest straight line

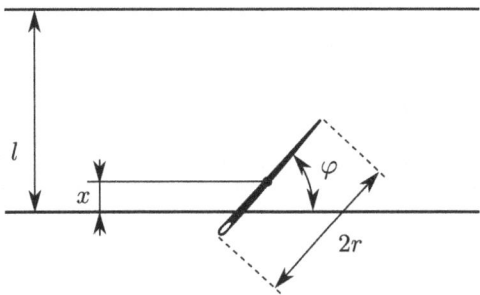

Fig. 2.2 Buffon's needle: the event in which the needle intersects one of the straight lines

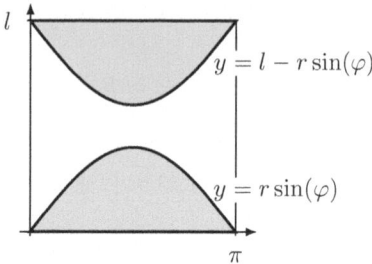

Let A be the event in which the needle intersects one of these straight lines and let $y = r \sin \varphi$. Then we get (see Fig. 2.2)

$$A = \{(x, \varphi) \in \Omega : x < r \sin \varphi \text{ or } x > l - r \sin \varphi\}.$$

Now it is easy to calculate $\lambda_2(A) = 2 \int_0^\pi r \sin \varphi \, d\varphi = 4r$, and $\mathbf{P}(A) = 4r/\pi l$.

A method of experimentally determining the number π is based on this task. Let's assume $l = 4r$. We throw the needle n times onto a plane lined with parallel lines at distance l apart and find the number $n(A)$ of those cases in which the needle touches one of the lines. In Chap. 7 on limit theorems we will prove that

$$\lim_{n \to \infty} \frac{n(A)}{n} = \frac{1}{\pi},$$

thus we have the approximation $\pi \approx n/n(A)$ for large n.

Remark 2.14 Many modifications of this paradox can be found on the Internet, in particular, studies called **Buffon's Noodle**. Please note that some of these studies contain obvious errors, which can be deduced from the following easily proven fact: For any small $\varepsilon > 0$ and any large $\ell > 0$, a ball of radius ε contains a polyline of length ℓ.

Example 2.15 (Bertrand's Paradox) We want to answer the following question: Suppose we have a circle of radius R. We choose a random chord of the circle, i.e. a line segment joining two points on the circle. What is the probability that this chord will be longer than the side of an equilateral triangle inscribed in the circle?

Bertrand noted that the question posed in this way does not clearly define what it means to *randomly* choose a chord of a circle. Figure 2.3 shows the circle $K(0, R)$, a triangle inscribed in the circle, and the circle $K(0, r)$ inscribed in the triangle. Obviously, $2r = R$. Let A be the event in which a randomly selected chord is longer than the side of an inscribed equilateral triangle. Now, let us consider how to *randomly* select the chord.

The First Way We can reason in the following way: since we are interested only in the length of the chord, we can assume that one of the ends of the chord is a fixed vertex of an inscribed equilateral triangle. The other end of the chord is any point of

2.6 Geometric Probability

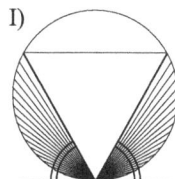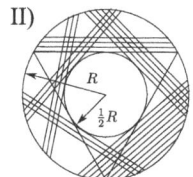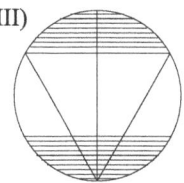

Fig. 2.3 Bertrand's paradox: choosing a random chord of a circle

the circle of circumference $2\pi R$ and the position of this second point describes the chord uniquely. We can assume now that $\Omega = [0, 2\pi R)$ and it seems to be natural that all the points from Ω have the same probability of being the other end of the chord. It is easy to see that the event A contains only those points of the circle that are between the other two vertices of the triangle. Hence, it follows that $\mathbf{P}(A) = 1/3$.

The Second Way It is known that the position of the center of a chord, except for a diameter, defines the chord uniquely—it is orthogonal to the line connecting the midpoint of the chord with the center of the circle. Moreover, each inner point of the ball, except its center, is the midpoint of some uniquely defined chord, thus we can assume that $\Omega = K(0, R)$, then $\lambda_2(\Omega) = \pi R^2$. The event A includes only those points of Ω that are inside the circle $K(0, R/2)$, hence $\lambda_2(A) = \pi R^2/4$ and, consequently, $\mathbf{P}(A) = 1/4$.

The Third Way The length of the chord is clearly determined by the distance between its center and the center of the circle. So we can assume that $\Omega = [0, R)$. Only the points in the segment $[0, R/2)$ belong to the event A. If we assume that all points of Ω are equally probable, then $\mathbf{P}(A) = 1/2$.

We have obtained three different solutions: $1/3$, $1/4$ and $1/2$. *Each of them is correct*, but each of them solves a slightly different problem. The differences lie in the different concepts of *randomness* in the individual cases. In Bertrand's example, it turns out that different interpretations of *randomness* lead to different results. They may not always lead to different results, however. Thus, Bertrand's paradox shows how important it is to strictly define the conditions of a random experiment in the stochastic modeling of real phenomena.

Remark 2.16 As in the case of putting objects into bins, physicists, when describing reality, choose from all possible interpretations of randomness those that describe reality best. Therefore, it is not surprising that some of them can argue that in the case of Bertrand's paradox "the third way" is, in some sense, more correct than the others. For example, see Jaynes [9]. However, mathematicians are obliged to consider all logically correct interpretations.

2.6.1 Exercises

84. What is the probability that a number x randomly chosen from the interval $[0, 5]$ satisfies the condition $(x - 1)^2 > 3$?
85. What is the probability that a number x randomly chosen from the interval $[0, 5]$ belongs to the domain of the function $f(x) = \log \sqrt{2 \cos x - 1}$?
86. Two friends are to meet between 10 a.m. and 11 a.m. They are coming to the appointed place independently of each other and each of them has promised to wait 20 minutes for the other. If the friend does not show up within this time, the one who is waiting will leave. What is the probability that they will meet?
87. We divide an interval of length ℓ into three parts at random. What is the probability that a triangle can be constructed from the obtained intervals?
88. Randomly throw a coin of radius R onto a plane lined with parallel lines at distance l apart, $2R < l$. Find the probability that the coin will not touch any of the lines.
89. From the interval $[-1, 1]$, we randomly choose two numbers p and q. Calculate the probability that the quadratic equation $x^2 + px + q = 0$ has two real roots.
90. We consecutively draw three numbers x_1, x_2, x_3 from the interval $[a, b]$. Calculate the probability that the third number x_3 will fall into the interval between the first two.
91. We randomly choose n points from the interval $[0, b]$. What is the probability that exactly k of these points will fall into the interval $[0, a]$, $k \leq n$, $a < b$?
92. Randomly pick n points from a ball with radius R. What is the probability that the distance of each of these points from the center of the sphere is not less than a, $0 < a < R$?
93. Let's randomly choose two points A and B from a square K. Calculate the probability that the square with diagonal AB is entirely contained in K.
94. We choose one point at random from a sphere with radius R. Calculate the probability that the distance of this point from a fixed diameter of the sphere is greater than a, $0 < a < R$.
95. We choose one point at random from a circle with radius R. Calculate the probability that the distance of this point from a fixed diameter of the circle is greater than a, $0 < a < R$.
96. We randomly select two points from the circle with radius R. Calculate the probability that their distance is less than x, where $x \in (0, 2R)$.

2.7 Conditional Probability

Definition 2.17 Let $(\Omega, \mathcal{F}, \mathbf{P})$ be probability space and let $B \in \mathcal{F}$ be a random event such that $\mathbf{P}(B) > 0$. For every random event $A \in \mathcal{F}$, we define

$$\mathbf{P}(A|B) \stackrel{\text{def}}{=} \frac{\mathbf{P}(A \cap B)}{\mathbf{P}(B)}.$$

2.7 Conditional Probability

The number $\mathbf{P}(A|B)$ is interpreted as the probability of the event A, provided that the event B has occurred, and the function

$$\mathbf{P}(\cdot|B) : \mathcal{F} \to [0, 1]$$

is called the *conditional probability* given B.

For example, consider throwing a die. If the event A means that we have rolled an even number and B means that we have rolled a number divisible by 3, then $\mathbf{P}(B) = 1/3$, $\mathbf{P}(A \cap B) = 1/6$, so $\mathbf{P}(A|B) = 1/2$. The obtained result agrees with the intuitive understanding of the concept *conditional probability, provided that the event B happened*. If you know that B has happened, you know that a three or a six has come out, and only one of these results meets the conditions of the event A, hence $\mathbf{P}(A|B) = 1/2$.

Theorem 2.18 (Law of Total Probability) *Let $B_1, B_2, \ldots, B_n$ be random events satisfying the following conditions:*

(1) $B_i \cap B_j = \emptyset$ for $i \neq j$;
(2) $\bigcup_{i=1}^{n} B_i = \Omega$;
(3) $\mathbf{P}(B_i) > 0$ for every $i = 1, 2, \ldots, n$.

Then, for every random event $A \in \mathcal{F}$, we have

$$\mathbf{P}(A) = \mathbf{P}(A|B_1)\mathbf{P}(B_1) + \mathbf{P}(A|B_2)\mathbf{P}(B_2) + \cdots + \mathbf{P}(A|B_n)\mathbf{P}(B_n).$$

Proof Since the sets $B_1, \ldots, B_n$ meet the assumptions (1) and (2), the event A can be written as a sum of pairwise disjoint sets:

$$A = A \cap \Omega = A \cap \bigcup_{i=1}^{n} B_i = \bigcup_{i=1}^{n} (A \cap B_i).$$

The probability of the sum of pairwise disjoint events is equal to the sum of the probabilities of these events, so finally, we get

$$\mathbf{P}(A) = \sum_{i=1}^{n} \mathbf{P}(A \cap B_i) = \sum_{i=1}^{n} \mathbf{P}(A|B_i)\mathbf{P}(B_i).$$

□

Example 2.19 Paul is playing an RPG. The wizard who is his playing figure will win a showdown if Paul rolls a single die and obtains a number divisible by three. The die must first be drawn from a box containing three dice: a cube (6 faces), an octahedron (8 faces), and a dodecahedron (12 faces). Each of these dice has faces numbered with consecutive natural numbers starting from 1. What is the probability that the wizard will win?

The result of throwing a die depends on the kind of die that is thrown. Let B_1 mean that Paul has chosen a cube, B_2 that he has chosen an octahedron and B_3 that he has chosen a dodecahedron. Since he is supposed to draw only one die, the events B_1, B_2 and B_3 are mutually exclusive and exhaust all possibilities. We assume that Paul is not previewing during the draw, so $\mathbf{P}(B_1) = \mathbf{P}(B_2) = \mathbf{P}(B_3) = 1/3 > 0$. Let A be the event of rolling a number that is divisible by three. Then $\mathbf{P}(A|B_1) = 1/3$, $\mathbf{P}(A|B_2) = 2/8$, $\mathbf{P}(A|B_3) = 4/12$. From the Law of Total Probability, we obtain

$$\mathbf{P}(A) = \frac{1}{3} \cdot \frac{1}{3} + \frac{2}{4} \cdot \frac{1}{3} + \frac{4}{12} \cdot \frac{1}{3} = \frac{11}{36}.$$

Example 2.20 (The Monty Hall Problem) In the recently popular TV game *Let's Make a Deal*, a player has three boxes to choose from. In one of them there is an attractive reward, while in the other two, there are cute but much less attractive black and white teddy cats, both named Zonk. Assume that the player has chosen box A. In the second stage of the game, the host informs the player that there is a Zonk in one of the remaining boxes, for example in C. The player can then either stick to his original choice and leave the game with the contents of box A, or change his mind and choose box B. Which decision is better?

At first, it seems that each of these decisions is equally good, but are they really? Let A mean there is a prize in box A, and A' mean there is a Zonk in box A. Similarly, we define B, B', C, C'. Let

$$B_1 = A \cap B' \cap C', \quad B_2 = A' \cap B \cap C', \quad B_3 = A' \cap B' \cap C.$$

Of course $\mathbf{P}(B_i) = 1/3$ for $i = 1, 2, 3$ and $\Omega = B_1 \cup B_2 \cup B_3$. It would seem that the probability of the prize being in box B, after receiving the information (the hint) from the host, is equal to the conditional probability provided by C', so it is equal to $1/2$. By reasoning in this way, we only use a fraction of the information we have. Yet, we know more—we know that the host has decided to show us box C. Depending on the situation, he either has had or has not had a choice! Hence, let D denote the event where the host has said that there is a Zonk in box C. We then have:

$$\mathbf{P}(B|D) = \frac{\mathbf{P}(B \cap D)}{\mathbf{P}(D)} = \frac{\mathbf{P}(B_2)}{\mathbf{P}(D)}.$$

It remains to calculate $\mathbf{P}(D)$. Note that if B_1 happened, the host would point to box C with probability $1/2$, and if B_2 happened, he would have no choice but to point to C with probability 1. Hence,

$$\mathbf{P}(D) = \mathbf{P}(D|B_1)\mathbf{P}(B_1) + \mathbf{P}(D|B_2)\mathbf{P}(B_2) + \mathbf{P}(D|B_3)\mathbf{P}(B_3)$$

$$= \frac{1}{2} \cdot \frac{1}{3} + 1 \cdot \frac{1}{3} + 0 \cdot \frac{1}{3} = \frac{1}{2}.$$

2.7 Conditional Probability

Consequently, in the end, $\mathbf{P}(B|D) = 2/3$, and it turns out that in this case it is worthwhile to change one's mind.

Theorem 2.21 (Bayes' Formula) *Assume that the random events $A, B_1, \ldots, B_n$ satisfy the following conditions:*

(1) $B_i \cap B_j = \emptyset$ for $i \neq j$;
(2) $\bigcup_{i=1}^{n} B_i = \Omega$;
(3) $\mathbf{P}(B_i) > 0$ for each $i = 1, 2, \ldots, n$;
(4) $\mathbf{P}(A) > 0$.

Then, for every $i = 1, 2, \ldots, n$, we have:

$$\mathbf{P}(B_i|A) = \frac{\mathbf{P}(A|B_i)\mathbf{P}(B_i)}{\mathbf{P}(A|B_1)\mathbf{P}(B_1) + \mathbf{P}(A|B_2)\mathbf{P}(B_2) + \cdots + \mathbf{P}(A|B_n)\mathbf{P}(B_n)}.$$

Proof It is enough to note that

$$\mathbf{P}(B_i|A) = \frac{\mathbf{P}(B_i \cap A)}{\mathbf{P}(A)} = \frac{\mathbf{P}(A|B_i)\mathbf{P}(B_i)}{\mathbf{P}(A)},$$

and then apply the Law of Total Probability. □

Let's go back to the example of Paul playing the RPG and suppose he has rolled a number divisible by 3. From Bayes' Formula, it is easy to calculate the probability that Paul had rolled an octagonal die:

$$\mathbf{P}(B_2|A) = \frac{\mathbf{P}(A|B_2)\mathbf{P}(B_2)}{\mathbf{P}(A)} = \frac{\frac{1}{4} \cdot \frac{1}{3}}{\frac{11}{36}} = \frac{3}{11}.$$

2.7.1 Exercises

97. Prove that the conditional probability satisfies the probability axioms.
98. Prove the Law of Total Probability and Bayes' Formula in the case of a countable partition of the set Ω, i.e., if $\Omega = \bigcup_{n=1}^{\infty} B_n$, $B_i \cap B_j = \emptyset$ for $i \neq j$, $\mathbf{P}(B_n) > 0$ for each $n \in \mathbb{N}$ and $\mathbf{P}(A) > 0$.
99. Prove that if $\mathbf{P}(A_1 \cap A_2 \cap \cdots \cap A_{n-1}) > 0$, then

$$\mathbf{P}(A_1 \cap A_2 \cap \cdots \cap A_n) = \mathbf{P}(A_1)\mathbf{P}(A_2|A_1) \cdots \mathbf{P}(A_n|A_1 \cap A_2 \cap \cdots \cap A_{n-1}).$$

100. Prove that if $B_1, \ldots, B_n$ are disjoint, $\mathbf{P}(B_i \cap C) > 0$ for every $i = 1, 2, \ldots, n$ and $A \cap C \subset \bigcup_{k=1}^{n} B_k$, then

$$\mathbf{P}(A|C) = \sum_{k=1}^{n} \mathbf{P}(A|B_k \cap C)\mathbf{P}(B_k|C).$$

101. Assume that $\mathbf{P}(A|B) > \mathbf{P}(B|A)$, $\mathbf{P}(A) > 0$ and $\mathbf{P}(B) > 0$. Is it true that $\mathbf{P}(A) > \mathbf{P}(B)$?
102. Is the following equality true: $\mathbf{P}(A|B) + \mathbf{P}(A|B') = 1$, where $B' = \Omega \setminus B$?
103. We have three urns containing white and black balls, with the ratio of the number of white balls to the number of black balls being α in the first, β in the second and γ in the third urn. We choose an urn randomly and then we draw a ball from it. What is the probability that it is white?
104. There are two shooters at a shooting range. The first one hits a target with a probability of 0.5, the second one of 0.8. They have tossed a coin to determine which of them will shoot. An outside observer, who can see the results but cannot see the shooters, observes that the shot has hit the target. What is the probability that it was the first shooter who fired?
105. Three balls were drawn randomly from an urn containing 7 white and 3 black balls. If it is known that a black ball is among those drawn, what is the probability that the other two are white?
106. An urn contains n white balls and m black balls. We draw one ball and then throw it back into the urn, adding ℓ white balls if it was a white ball or ℓ black balls if it was black. We repeat this operation many times. Prove that the probability of getting a white ball in the k-th step is equal to $\frac{n}{n+m}$ for: (a) $k = 1, 2, 3$; (b) any k.
107. There are n balls in an urn, including m white ones. We randomly draw k balls. Denote by A_i the event where the i-th drawn ball is white, and by B_j the event where the white ball is drawn j times. Prove that $\mathbf{P}(A_i|B_j) = j/k$ for both 'with replacement' and 'without replacement' randomization.
108. Eugene had N 5-cent and M 10-cent coins in his wallet, but he lost a coin and doesn't know what denomination it was. Two coins drawn randomly from the wallet turned out to be 5-cent coins. What is the probability that the lost coin was a 10-cent coin?
109. To make herself save money, Karen puts every ten- and twenty-cent coin she finds in her wallet into a piggy bank at the end of every day. After emptying the bank last time, one coin remained in it, and today Karen added a 10-cent coin. A coin pulled randomly from the bank turned out to be a 10-cent coin. What is the probability that there is another 10-cent coin in the box?
110. In a group of 30 students, there are five students who always pass an exam with an A-grade. There are also ten students who always get an A- or B-grade in an examination with the same probability. The remaining fifteen receive a B, C, or F with equal probability. What is the probability that a randomly selected student from this group will receive: (a) A; (b) B?
111. We have 3 urns. In the first, there are 3 black and 7 white balls; in the second, 4 black and 6 white balls; in the third 6 black and 4 white balls. We randomly take one ball from the first urn and put it into the second. Then, we take a random ball from the second urn and put it into the third. What is now the probability that a ball randomly taken from the third urn is white?

112. Nuts from three hazel bushes have been harvested and placed in three baskets so that each basket has nuts from a different bush, but we don't know which bush the contents of a given basket comes from. Two of these shrubs have not received any horticultural treatment and 1/3 of their nuts are infested with worms. On the third shrub, only 5% of the nuts are infested. Is it possible to find a strategy for buying 3 kgs of nuts so that with a probability of more than 1/2 you will buy less than 0.7 kgs of worm-eaten nuts?

2.8 Independent Events

Independence of random events is a very important idea in probability theory. It can be defined by using conditional probability and claiming that A and B are independent if the conditional probability $\mathbf{P}(A|B)$ does not depend on the condition. This, however, unjustly excludes the situation when $\mathbf{P}(B) = 0$, thus we shall use the following definition:

Definition 2.22 We say that random events A, B are *independent* if

$$\mathbf{P}(A \cap B) = \mathbf{P}(A)\mathbf{P}(B).$$

If for a sequence of random events $A_1, A_2, \ldots, A_n$ the following condition holds

$$\forall i, j = 1, \ldots, n, \ i \neq j, \quad \mathbf{P}(A_i \cap A_j) = \mathbf{P}(A_i)\mathbf{P}(A_j),$$

we say that the events $A_1, A_2, \ldots, A_n$ are *pairwise independent*.

Definition 2.23 We say that random events $A_1, A_2, \ldots, A_n$ are *independent* if for every $k \leq n$ and any choice of different indexes $n_1, \ldots, n_k \in \{1, 2, \ldots, n\}$, the following condition holds:

$$\mathbf{P}(A_{n_1} \cap \cdots \cap A_{n_k}) = \mathbf{P}(A_{n_1}) \ldots \mathbf{P}(A_{n_k}).$$

So, to prove that A, B and C are independent, we need to check if they are pairwise independent, and that $\mathbf{P}(A \cap B \cap C) = \mathbf{P}(A)\mathbf{P}(B)\mathbf{P}(C)$. If the events A, B and C are pairwise independent, it does not mean that they are independent!

2.8.1 Exercises

113. Prove that if the events A and B are independent, then the events A' and B' are independent and the events A and B' are independent.
114. Can an event A be independent of itself?

115. Assume that $\mathbf{P}(A) > 0$ and $\mathbf{P}(B) > 0$. Prove that

    (a) if A and B are disjoint then they are not independent;
    (b) if A and B are independent then they are not disjoint.

116. Show that if $\mathbf{P}(A|B) = \mathbf{P}(A|B')$, then the events A and B are independent.

117. The probability of event A is equal to 0 or 1. Prove that this event is independent of any event B.

118. Events A and B are independent and such that $\mathbf{P}(A \cup B) = 1$. Prove that $\mathbf{P}(A) = 1$ or $\mathbf{P}(B) = 1$.

119. Let A and B be independent events. Prove that if $A \cup B$ and $A \cap B$ are independent, then $\mathbf{P}(A) = 0$ or $\mathbf{P}(A) = 1$, or $\mathbf{P}(B) = 0$, or $\mathbf{P}(B) = 1$.

120. We roll a single die three times. The event A is when we obtain the same result on the first and second throw; event B - we got the same number on the second and third throw; C—on the first and third throw. Are the events A, B and C independent? Are they pairwise independent?

121. Events A and B are independent, and C is independent of $A \cup B$ and $A \cap B$. Can the events A, B and C be dependent?

122. Events A, B and C are pairwise independent and have probabilities that are not equal to zero or one. Can the events $A \cap B$, $B \cap C$ and $A \cap C$ be:
    (a) pairwise independent; (b) independent?

123. Show that the equality $\mathbf{P}(A \cap B \cap C) = \mathbf{P}(A)\mathbf{P}(B)\mathbf{P}(C)$ does not imply that the events A, B and C are independent.

124. We roll a single die twice. Event A holds if we got a number divisible by 3 on the first throw; B if the sum of obtained numbers is even; C if we got the same number on each throw. Are the events A, B and C independent? Are they pairwise independent?

125. Three students prepared independently for an exam in probability calculus. The probabilities of passing the exam for each student are: $p_1 = 0.6$; $p_2 = 0.5$ and $p_3 = 0.4$. Find the probability that the third student has passed the exam if we know that only two of them have passed.

126. At most, how many conditions need to be checked to prove the independence of the events $A_1, \ldots, A_n$? How many counterexamples need to be found to show that all of these conditions are relevant?

2.9 Bernoulli Trials

Consider a random experiment that consists of a series of n trials, where

(1) subsequent attempts are independent;
(2) in each trial, two outcomes are possible: one, called *success*, occurs with probability p; the other, called *failure*, has a probability of $q = 1 - p$.

2.9 Bernoulli Trials

Such trials are called *Bernoulli trials* and a sequence of Bernoulli trials is called a *Bernoulli experiment* with parameters n and p. The probability that in n trials we obtain k successes and $n - k$ failures is equal to:

$$\mathbf{P}\{X_n = k\} = \binom{n}{k} p^k (1 - p)^{n-k}, \qquad k = 0, 1, \ldots, n.$$

Justification of the Formula The result of a Bernoulli experiment can be described as a sequence of n elements, writing a one in the i-th place if we have had a "success" in the i-th trial; and writing a zero if we have failed in the i-th trial. Then, the record

$$\underbrace{1, \ldots, 1}_{k \text{ times}}, \underbrace{0, \ldots, 0}_{n-k \text{ times}}$$

means that we have had successes in the first k attempts, and failures in the remaining ones. The probability of such a result is equal to $p^k q^{n-k}$. We should also note that the number of n element sequences of k ones and $(n - k)$ zeros is equal to $\binom{n}{k}$.

Example 2.24 We toss a coin ten times. The probability of getting exactly 5 heads is equal to

$$\mathbf{P}\{X_{10} = 5\} = \binom{10}{5} \left(\frac{1}{2}\right)^{10} = \frac{63}{256} \approx 0.246094.$$

Example 2.25 We roll a single die 10 times and we would like to see how many sixes we get. Thus, getting a six means a success in this experiment, i.e., $p = \frac{1}{6}$ and, e.g.,

$$\mathbf{P}\{X_{10} = 5\} = \binom{10}{5} \left(\frac{1}{6}\right)^5 \left(\frac{5}{6}\right)^5 = \frac{7 \cdot 5^5}{6^8} \approx 0.0130238.$$

2.9.1 Most Probable Number of Successes

Figure 2.4 shows that the probability of getting exactly k successes in a sequence of n Bernoulli trials is a function of k which grows at the beginning, and decreases towards the end.

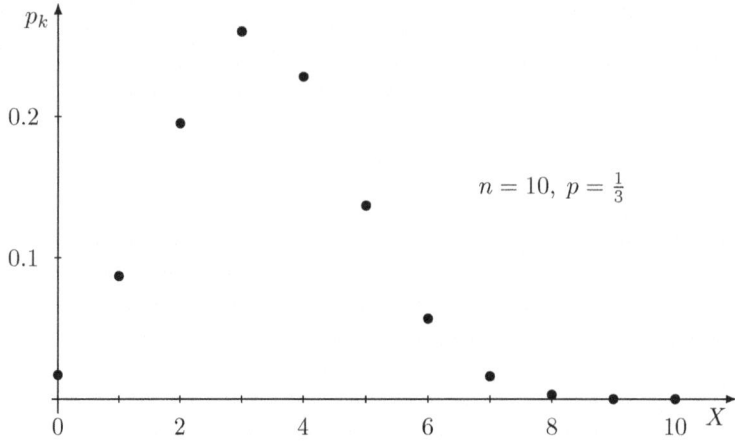

Fig. 2.4 Probability of getting k successes in a sequence of n Bernoulli trials

Intuitively, it is also clear that within 60 dice rolls, you are most likely to get 10 sixes, and within 60 coin flips you are most likely to get 30 heads. To justify this fact, consider the following expression:

$$A(k) = \frac{\mathbf{P}\{X_n = k+1\}}{\mathbf{P}\{X_n = k\}} = \frac{p(n-k)}{(1-p)(k+1)}.$$

It is easy to check that $\mathbf{P}\{X_n = k+1\} > \mathbf{P}\{X_n = k\}$ and consequently $A(k) > 1$ if $k < np + p - 1$ and $A(k) < 1$ if $k > np + p - 1$. Hence, we get that the most probable number of successes in n Bernoulli trials is any integer in the interval $[p(n+1) - 1, p(n+1)]$. If $p(n+1)$ is an integer, then there are two integers in this interval. Otherwise, the most likely number of successes is $[p(n+1)]$, where $[x]$ is the integer part of the number x.

Approximate Formulas Often, when solving problems involving a Bernoulli scheme, we will have to raise a number close to one to a very large power. It is worth remembering two formulas that facilitate approximate calculations here. One of them is based on Taylor's theorem, and we apply it when nx is a small number:

$$(1-x)^n = 1 - nx + o(nx).$$

The other formula is based on the well-known approximation of the constant e:

$$\lim_{n \to \infty} \left(1 - \frac{1}{n}\right)^n = e^{-1} \approx 0.36787944117\ldots$$

2.9 Bernoulli Trials

Example 2.26 Throughout his adult life, Mister Kowalski fills out a lottery coupon twice a week, selecting 6 out of 49 numbers. What is the probability that he will never hit all the winning six numbers?

First, let's calculate the probability of success, i.e., the probability of hitting *the six* in one draw:

$$\Omega = \{\{a_1, \ldots, a_6\} : a_i \in \{1, \ldots, 49\}, a_i \neq a_j \text{ for } i \neq j\};$$

$$n(\Omega) = \binom{49}{6} = 11 \cdot 12 \cdot 46 \cdot 47 \cdot 49.$$

Event A consists in selecting exactly all six randomly selected numbers

$$p = \mathbf{P}(A) = \frac{1}{n(\Omega)} = \frac{1}{11 \cdot 12 \cdot 46 \cdot 47 \cdot 49}.$$

Suppose Mr Kowalski's adult life will last at least 66 years. Buying two coupons a week for 66 years means that Mr Kowalski will make $2 \cdot 52 \cdot 66$ attempts to succeed. Hence, the probability that he will never hit *the six* is:

$$\mathbf{P}\{X_n = 0\} = \binom{n}{0} p^0 (1-p)^n = \left(1 - \frac{1}{11 \cdot 12 \cdot 46 \cdot 47 \cdot 49}\right)^{2 \cdot 52 \cdot 66}.$$

Using the first-order approximation from Taylor's formula, we get that

$$\mathbf{P}\{X_n = 0\} \approx 1 - \frac{2 \cdot 52 \cdot 66}{11 \cdot 12 \cdot 46 \cdot 47 \cdot 49} = 1 - \frac{26}{23 \cdot 47 \cdot 49} \approx 0.999509146 \ldots$$

On the other hand, using the approximation of e, we get

$$\mathbf{P}\{X_n = 0\} =$$

$$\left[\left(1 - \frac{1}{11 \cdot 12 \cdot 46 \cdot 47 \cdot 49}\right)^{11 \cdot 12 \cdot 46 \cdot 47 \cdot 49}\right]^{0.000490853} \approx 0.999509267 \ldots$$

Stirling's Approximate Formula Another important formula for approximate calculations involving the Bernoulli distribution and other distributions where the symbol $n!$ occurs is *Stirling's formula*:

$$n! \sim \sqrt{2\pi n} \cdot n^n \mathrm{e}^{-n},$$

where $\sim$ indicates that the quotient of the expressions on the left and right sides tends to one as n goes to infinity. The first proof of this formula was presented by James Stirling in 1730 [16]. Those interested can find this proof, for example, in the book by Feller [6], Sect. II.9, or in the book by Jakubowski and Sztencel [8], Sect. A.3.

2.9.2 The Black and White Hats Puzzle

There is a whole series of logical problems in which the most important element is guessing the colors of the hats that appear on some people's heads without them knowing which color they are wearing. Sometimes, when the color of the hat is not important, the tasks concern horns growing on some gentlemen's heads.

The simplest example of such a problem can be summarized as follows (see Fig. 2.5): three prisoners stand in a line, one after another, so that each of them can see only the prisoners standing in front of him, and the first cannot see anyone. They each have a black or white hat on their heads and it is known that only two hats are of the same color. The prisoners will be released if at least one of them can guess the color of his own hat. Will they be released?

We leave it to the reader to solve this puzzle, which only consists in carrying out precise logical reasoning. To make it easier, let us note that if the third prisoner sees two black hats in front of him, he knows the color of his hat, but if he sees hats of different colors, he should not try to answer the question.

In the 1990s, an interesting problem emerged which was of a seemingly similar type, but it was a probability problem, not a logic problem. Thanks to the Internet, it quickly reached many mathematicians and enthusiasts all over the world. It delighted with the simplicity of its formulation, but also the deep and unexpected relationships it has with many unsolved mathematical problems, including applications in telecommunications, computer science and coding theory.

In the *Black and White Hats puzzle*, a team of n players enters a room and a white or black hat is randomly placed on the head of each player. Each player can see the hats of all of the other players but not his own. The rules of the game exclude any form of communication between players. However, they may, before starting the game and entering the room, establish a common strategy. After seeing the hats of the other players, each of them can guess the color of their own hat or give up guessing. The team will win 3 million euros to be shared between them if at least one of the players guesses correctly and no other gives the wrong answer; otherwise

Fig. 2.5 Three prisoners

2.9 Bernoulli Trials

the team loses. The problem is to find a strategy that maximizes the probability of winning.

One obvious strategy guarantees a winning probability equal to $\frac{1}{2}$. It is enough to agree that one of the players, regardless of the situation, will declare that he or she has a black hat, and the others will be silent. However, is it possible to find a better strategy and increase this probability?

Note that for three players, the probability that they all have hats of the same color is $\frac{1}{4}$, and the probability that two hats are of one color and the third of the other is $\frac{3}{4}$, so it is worth assuming that we are in the latter situation. Let's agree that the player who can see two hats of the same color is to report that his or her hat is of the second color, and that the other players are silent. If there were two black hats and one white hat, only the player in the white hat speaks, claiming that he has a white hat and the group wins. However, if there were three black hats, which is three times less likely, each player would announce that he has a white hat and all of them would contribute equally to the loss. As you can see, this strategy has additional advantages! It favors teamwork: you stay silent if you presume that someone else has better information. What is more, it equally distributes the responsibility in the case of failure.

If the number of players is greater, the task becomes more complicated. However, you can always find a strategy that will lead to a win in most cases, and a less frequent loss. We suggest the reader to find a good strategy for $n = 7$. It is known that when the number of players is one less than a power of two, i.e., $n = 2^k - 1$, there is a strategy with a probability of winning equal to $1 - 2^{-k}$.

The optimal strategy you are looking for has a very elegant description in the language of Hamming codes, named after its author, mathematician Richard Hamming. These codes are used to remove errors in data transmission by all types of electronic devices, from telephone exchanges to optical discs. They are also used for data compression in computer memory.

2.9.3 Exercises

127. A series of lectures consists of 15 independent topics discussed in separate lectures. At the beginning of the semester, a special committee prepares a list of 5 examination tasks, each of them on a different topic. In order to pass the exam, the student has to solve three or more problems. The lecturer comes to the lecture with a probability of 0.92, and if he does not, the topic for that lecture is never discussed. What are the chances that Adam can pass the exam if he can solve the problems related to the topics discussed with a probability of 0.8, and cannot solve the topics that have not been covered?
128. We throw a coin n times, $n \geqslant 2$. What is the chance that heads will appear an even number of times?

129. We throw a single die n times. Calculate the probability that

   (a) the number 6 will appear exactly once;
   (b) the number 6 will appear at least once.

130. Which of the following events is more likely: A—a six will appear at least once in four dice rolls; B—in 24 rolls of two dice at least one of the rolls will show a pair of sixes?

131. We roll a single die several times. Which of the following events is more probable: A—the sum of the numbers rolled is even; or B—the sum is odd?

132. Two players each toss a coin n times. What is the probability that they obtain the same number of heads?

133. **Banach's Problem.** The Polish mathematician Stefan Banach not so much posed as inspired the following problem: A man had two boxes of matches, n matches in each, and he put one in his right and the other in his left jacket pocket. Each time he needed a match, he would reach randomly into one of his pockets. When he reached into his pocket again, it turned out that the box he pulled out was empty. What is the probability that the second box at that moment contained exactly k matches, $k \leqslant n$?

    Hint: You need to consider the case where the box in the right pocket is empty, and the case in which the box in the left pocket is empty, bearing in mind that both boxes may turn out to be empty at the same time.

134. The probability of drawing a winning lottery ticket is 0.25. How many tickets do I need to purchase in order to win with a probability of at least 0.9?

135. What is the probability that Mr Kowalski will not even hit a four (i.e. match four numbers) by playing a lottery twice a week for a year (when choosing 6 numbers out of 49)?

136. The probability of hitting a target with a single shot is p, and the probability of destroying the target with k hits, $k \geq 0$, is $1 - \lambda^k$. What is the probability of destroying the target if n shots are fired?

    Hint. Apply the total probability formula for $\{X_n = k\}, k = 0, 1, \ldots, n$, where X_n is the number of shots on target among n fired.

137. Find the optimal strategy for $n = 4, 5, 6$ players in the *Black and White Hats puzzle*.

138. Find the optimal strategy for $n = 7$ players.

2.10 Upper and Lower Limits of Sequences of Events

Let $(\Omega, \mathcal{F}, \mathbf{P})$ be a probability space and let $A_n \in \mathcal{F}, n \in \mathbb{N}$.

2.10 Upper and Lower Limits of Sequences of Events

Definition 2.27 The *upper limit* of a sequence of events A_n, $n \in \mathbb{N}$, is the set of $\omega \in \Omega$ which belong to infinitely many events A_n, i.e.,

$$\limsup_{n\to\infty} A_n \stackrel{\text{def}}{=} \{\omega : \forall n \geq 1 \, \exists m \geq n \, \omega \in A_m\} = \bigcap_{n=1}^{\infty} \bigcup_{m=n}^{\infty} A_m.$$

The *lower limit* of the sequence of events A_n, $n \in \mathbb{N}$, is the set of $\omega \in \Omega$ which belong to almost all events A_n, i.e., all except a finite number of events A_n:

$$\liminf_{n\to\infty} A_n \stackrel{\text{def}}{=} \{\omega : \exists n \geq 1 \, \forall m \geq n \, \omega \in A_m\} = \bigcup_{n=1}^{\infty} \bigcap_{m=n}^{\infty} A_m.$$

It is easy to see that both $\liminf_{n\to\infty} A_n$ and $\limsup_{n\to\infty} A_n$ are random events because the σ-field $\mathcal{F}$ is closed under countable set operations. Of course:

$$\liminf_{n\to\infty} A_n \subset \limsup_{n\to\infty} A_n.$$

The following theorem, which describes the basic properties of the upper and lower limits of a sequence of sets, is a special case of Fatou's theorem.

Theorem 2.28 *If (A_n) is a sequence of random events, then*

$$\mathbf{P}\left(\liminf_{n\to\infty} A_n\right) \leq \liminf_{n\to\infty} \mathbf{P}(A_n) \leq \limsup_{n\to\infty} \mathbf{P}(A_n) \leq \mathbf{P}\left(\limsup_{n\to\infty} A_n\right).$$

Proof Let $B_n = \bigcap_{k=n}^{\infty} A_k$ and $C_n = \bigcup_{k=n}^{\infty} A_k$. The sequences (B_n) and (C_n) are, respectively, increasing and decreasing sequences of sets. In addition,

$$\liminf_{n\to\infty} A_n = \bigcup_{n=1}^{\infty} B_n, \qquad \limsup_{n\to\infty} A_n = \bigcap_{n=1}^{\infty} C_n.$$

By Property 8 of the probability measure **P** (Continuity Theorem), we get:

$$\mathbf{P}(A_n) \geq \mathbf{P}(B_n) \longrightarrow \mathbf{P}\left(\liminf_{n\to\infty} A_n\right) \text{ and}$$

$$\mathbf{P}(A_n) \leq \mathbf{P}(C_n) \longrightarrow \mathbf{P}\left(\limsup_{n\to\infty} A_n\right),$$

from which the result easily follows. □

Lemma 2.29 (Borel–Cantelli Lemma) *Let $(\Omega, \mathcal{F}, \mathbf{P})$ be a probability space and let $A_n \in \mathcal{F}$ for each $n \in \mathbb{N}$.*

(1) *If $\sum_{n=1}^{\infty} \mathbf{P}(A_n) < \infty$, then $\mathbf{P}\left(\limsup_{n \to \infty} A_n\right) = 0$.*
(2) *If the events A_n are independent and such that $\sum_{n=1}^{\infty} \mathbf{P}(A_n) = \infty$, then $\mathbf{P}\left(\limsup_{n \to \infty} A_n\right) = 1$.*

Proof The proof of (1) is simple:

$$\mathbf{P}\left(\limsup_{n \to \infty} A_n\right) \leqslant \mathbf{P}\left(\bigcup_{k=n}^{\infty} A_k\right) \leqslant \sum_{k=n}^{\infty} \mathbf{P}(A_k) \longrightarrow 0,$$

which is due to the convergence of the series $\sum_{n=1}^{\infty} \mathbf{P}(A_n)$. Assume now that the events A_n are independent. Then, the A'_n are independent too, and

$$\mathbf{P}\left(\Omega \setminus \bigcup_{k=n}^{\infty} A_k\right) = \mathbf{P}\left(\bigcap_{k=n}^{\infty} (\Omega \setminus A_k)\right) = \prod_{k=n}^{\infty} (1 - \mathbf{P}(A_k))$$

$$\leqslant \prod_{k=n}^{\infty} e^{-\mathbf{P}(A_k)} = \exp\left\{-\sum_{k=n}^{\infty} \mathbf{P}(A_k)\right\} = 0,$$

since $\sum_{k=n}^{\infty} \mathbf{P}(A_k) = \infty$. This implies that $\mathbf{P}\left(\bigcup_{k=n}^{\infty} A_k\right) = 1$ for each $n \in \mathbb{N}$. Since the events $\bigcup_{k=n}^{\infty} A_k$ form a decreasing sequence, we get

$$\mathbf{P}\left(\limsup_{n \to \infty} A_n\right) = \lim_{n \to \infty} \mathbf{P}\left(\bigcup_{k=n}^{\infty} A_k\right) = 1.$$

$\square$

2.10.1 Exercises

139. Show that for any sequence of random events $A_n, n \in \mathbb{N}$

$$\left(\limsup_{n \to \infty} A_n\right)' = \liminf_{n \to \infty} A'_n, \quad \left(\liminf_{n \to \infty} A_n\right)' = \limsup_{n \to \infty} A'_n.$$

140. Let $A_k, k \in \mathbb{N}$, be a sequence of random events in a fixed probability space $(\Omega, \mathcal{F}, \mathbf{P})$. Prove that the random events $\liminf A_n$ and $\limsup A_n$ belong to the σ-field $\sigma(\{A_k : k \in \mathbb{N}\})$.

2.10 Upper and Lower Limits of Sequences of Events

141. Are the following relationships true:

$$\mathbf{P}\left(\limsup_{n\to\infty} A_n\right) = \lim_{n\to\infty} \mathbf{P}\left(\bigcup_{k=n}^{\infty} A_k\right); \quad \mathbf{P}\left(\liminf_{n\to\infty} A_n\right) = \lim_{n\to\infty} \mathbf{P}\left(\bigcap_{k=n}^{\infty} A_k\right)?$$

142. Prove that for any $A_n, B_n \in \mathcal{F}, n \in \mathbb{N}$

$$\limsup_{n\to\infty}(A_n \cup B_n) = \left(\limsup_{n\to\infty} A_n\right) \cup \left(\limsup_{n\to\infty} B_n\right);$$
$$\liminf_{n\to\infty}(A_n \cap B_n) = \left(\liminf_{n\to\infty} A_n\right) \cap \left(\liminf_{n\to\infty} B_n\right).$$

143. What is the probability that in an infinite sequence of tosses of a symmetric coin, heads will appear finitely many times?
144. What is the probability that a six will appear infinitely many times in an infinite series of dice rolls?

Chapter 3
Random Variables and Their Distributions

3.1 Definition of a Random Variable

As we have already seen, the description of a probability space can be complicated and troublesome in some cases. Usually, however, we are not interested in the result of a random experiment itself, but in the value of a certain function ascribed to this event. This could be, for example, the value of the sum of the results of several dice rolled at the same time, the value of a win (or loss) in a lottery, or, in a sequence of consecutive coin tosses, which one of them will be the first to fall heads up.

If Ω consists of a finite number of elements, then each function $X : \Omega \to \mathbb{R}$ has the property that we can calculate the probability of $X^{-1}(B)$ for any Borel set $B \in \mathcal{F}$. However, if Ω is uncountable, or if $\mathcal{F}$ does not contain all subsets of the countable set Ω, then we can only consider functions X for which the probability of $X^{-1}(B)$ is well defined for all Borel sets $B \subset \mathbb{R}$. We will call the functions that satisfy this condition *random variables*.

Definition 3.1 A function X defined on the probability space $(\Omega, \mathcal{F}, \mathbf{P})$ taking values in $\mathbb{R}$ is called a *random variable* if for every $t \in \mathbb{R}$, the following condition holds:

$$X^{-1}((-\infty, t)) = \left\{\omega \in \Omega : X(\omega) < t\right\} \in \mathcal{F}.$$

The next two theorems give equivalent definitions of random variables.

Theorem 3.2 *Let* $X : (\Omega, \mathcal{F}, \mathbf{P}) \to \mathbb{R}$. *Then, the following conditions are equivalent:*

(1) *X is a random variable;*
(2) *for every $t \in \mathbb{R}$, we have $\{\omega : X(\omega) \leqslant t\} \in \mathcal{F}$;*
(3) *for every $t \in \mathbb{R}$, we have $\{\omega : X(\omega) > t\} \in \mathcal{F}$;*
(4) *for every $t \in \mathbb{R}$, we have $\{\omega : X(\omega) \geqslant t\} \in \mathcal{F}$.*

Proof Note that $\mathcal{F}$ is a σ-field, hence it is closed under countable set operations. The equivalence of (1) $\Leftrightarrow$ (2) can therefore be easily deduced from the following facts:

$$\{\omega : X(\omega) \leqslant t\} = \bigcap_{n \in \mathbb{N}} \{\omega : X(\omega) < t + 1/n\};$$
$$\{\omega : X(\omega) < t\} = \bigcup_{n \in \mathbb{N}} \{\omega : X(\omega) \leqslant t - 1/n\}.$$

Similarly, we show that (3) and (4) are equivalent. To show that (1) implies 4) note that

$$\{\omega : X(\omega) \geqslant t\} = \Omega \setminus \{\omega : X(\omega) < t\} \in \mathcal{F}.$$

Similarly, condition (4) implies (1) because

$$\{\omega : X(\omega) < t\} = \Omega \setminus \{\omega : X(\omega) \geqslant t\} \in \mathcal{F},$$

which was to be shown. □

Theorem 3.3 *The function* $X : (\Omega, \mathcal{F}, \mathbf{P}) \to (\mathbb{R}, \mathcal{B})$ *is a random variable if and only if for any Borel set* $B \in \mathcal{B}$, *the following condition holds:*

$$\{\omega : X(\omega) \in B\} \in \mathcal{F}. \qquad (*)$$

Proof Of course, if the condition $(*)$ is true, then X is a random variable, because the half-lines $(-\infty, t), t \in \mathbb{R}$, are Borel sets. Hence, let's assume that X is a random variable and denote by $\mathcal{K}$ the family of all Borel subsets of the line satisfying the condition $(*)$, i.e.,

$$\mathcal{K} \stackrel{\text{def}}{=} \{B \in \mathcal{B} : \{\omega : X(\omega) \in B\} \in \mathcal{F}\}.$$

Since X is a random variable, for every $t \in \mathbb{R}$ the half-line $(-\infty, t)$ belongs to $\mathcal{K}$. It follows from the previous theorem that for every $t \in \mathbb{R}$ the set $(-\infty, t]$ belongs to $\mathcal{K}$. Consequently, all open intervals belong to $\mathcal{K}$ because for all $a < b$,

$$\{\omega : a < X(\omega) < b\} = \{\omega : X(\omega) < b\} \setminus \{\omega : X(\omega) \leqslant a\} \in \mathcal{F}.$$

Every open set $U \subset \mathbb{R}$ is a sum of a countably many open intervals $I_n, n \in \mathbb{N}$, thus $U \in \mathcal{K}$ because

$$\left\{\omega : X(\omega) \in \bigcup_n I_n\right\} = \bigcup_n \{\omega : X(\omega) \in I_n\} \in \mathcal{F}.$$

3.1 Definition of a Random Variable

It remains to prove that $\mathcal{K}$ is a σ-field, i.e., it is closed with respect to all countable set operations. Assume that for each $n \in \mathbb{N}$, we have $A_n \in \mathcal{K}$. Then,

$$\left\{\omega : X(\omega) \in A_1'\right\} = \Omega \setminus \{\omega : X(\omega) \in A_1\} \in \mathcal{F},$$

$$\left\{\omega : X(\omega) \in \bigcup_n A_n\right\} = \bigcup_n \{\omega : X(\omega) \in A_n\} \in \mathcal{F},$$

$$\left\{\omega : X(\omega) \in \bigcap_n A_n\right\} = \bigcap_n \{\omega : X(\omega) \in A_n\} \in \mathcal{F}.$$

We see now that $\mathcal{K}$ is a σ-field containing all open sets, i.e., containing the whole Borel σ-field $\mathcal{B}$. However, by our assumption $\mathcal{K}$ is a subset of the Borel σ-field, thus finally, $\mathcal{K} = \mathcal{B}$. □

Recall that a function $f : (\Omega, \mathcal{F}) \to (\mathbb{R}, \mathcal{B})$ is *measurable* if $\{\omega : f(\omega) \in B\} \in \mathcal{F}$ for any Borel set $B \in \mathcal{B}$. Hence, we conclude that each random variable is a measurable function from $(\Omega, \mathcal{F})$ to $(\mathbb{R}, \mathcal{B})$, but not every measurable function is a random variable! For this to be the case, the space $(\Omega, \mathcal{F})$ must be "equipped" with a probability measure.

Theorem 3.4 *If X and Y are random variables defined on the same probability space $(\Omega, \mathcal{F}, \mathbf{P})$, then the following functions are also random variables:*

(a) $aX(\omega)$, where a is a real number;
(b) $X(\omega) + Y(\omega)$;
(c) $X(\omega) \cdot Y(\omega)$.

Proof

(a) If $a = 0$, then the set $\{\omega : aX(\omega) < t\}$ is equal to either the empty set or Ω. Both of these sets belong to $\mathcal{F}$, hence $aX(\omega)$ is a random variable. If $a \neq 0$, then

$$\{\omega : aX(\omega) < t\} = \begin{cases} \{\omega : X(\omega) < t/a\} & \text{if } a > 0; \\ \{\omega : X(\omega) > t/a\} & \text{if } a < 0. \end{cases}$$

In both cases, the obtained sets belong to $\mathcal{F}$, thus the function $aX(\omega)$ is a random variable.

(b) To prove that $X(\omega) + Y(\omega)$ is a random variable, let us note that

$$\{\omega : X(\omega) + Y(\omega) < t\} = \{\omega : X(\omega) < t - Y(\omega)\}$$
$$= \bigcup_{q \in Q} \{\omega : X(\omega) < q, \, q < t - Y(\omega)\},$$

where Q denotes the set of rational numbers. Hence, it already follows that

$$\{\omega : X(\omega) + Y(\omega) < t\} = \bigcup_{q \in Q} \{\omega : X(\omega) < q\} \cap \{\omega : q < t - Y(\omega)\}$$

$$= \bigcup_{q \in Q} \{\omega : X(\omega) < q\} \cap \{\omega : Y(\omega) < t - q\} \in \mathcal{F}.$$

To get the final inclusion of sets, we have used the fact that the set of rational numbers is countable and dense in $\mathbb{R}$, and the fact that the σ-field $\mathcal{F}$ is closed with respect to countable operations.

(c) We will first show that if X is a random variable, then X^2 is also a random variable. For $t \leqslant 0$, we get $\{\omega : X^2(\omega) < t\} = \emptyset \in \mathcal{F}$; and if $t > 0$, then we conclude that

$$\{\omega : X^2(\omega) < t\} = \{\omega : X(\omega) < \sqrt{t}\} \setminus \{\omega : X(\omega) \leqslant -\sqrt{t}\} \in \mathcal{F}.$$

Now, we can show that $X(\omega)Y(\omega)$ is a random variable if $X(\omega)$ and $Y(\omega)$ are random variables. The previous considerations show that $X(\omega) + Y(\omega)$ and $X(\omega) - Y(\omega)$ are random variables. We conclude from this that $(X(\omega) + Y(\omega))^2$ and $(X(\omega) - Y(\omega))^2$ are random variables, thus also

$$X(\omega)Y(\omega) = \frac{1}{4}\left((X(\omega) + Y(\omega))^2 - (X(\omega) - Y(\omega))^2\right)$$

is a random variable.

□

Theorem 3.5 *Assume that (X_n) is a sequence of random variables such that for every fixed $\omega \in \Omega$*

$$\sup_n X_n(\omega) < \infty \qquad \left(\inf_n X_n(\omega) > -\infty\right).$$

Then, the function $X(\omega) = \sup_n X_n(\omega)$ (and $Y(\omega) = \inf_n X_n(\omega)$ respectively) is also a random variable.

Proof Let $t \in \mathbb{R}$. Both conclusions are easily derived from Theorem 3.2 and the following facts:

$$\left\{\omega : \sup_n X_n(\omega) > t\right\} = \bigcup_n \{\omega : X_n(\omega) > t\} \in \mathcal{F};$$

$$\left\{\omega : \inf_n X_n(\omega) < t\right\} = \bigcup_n \{\omega : X_n(\omega) < t\} \in \mathcal{F}.$$

□

3.1 Definition of a Random Variable

Corollary 3.6 *Let $X_n(\omega)$ be a sequence of random variables. If for each $\omega \in \Omega$, the condition $\sup_n X_n(\omega) < \infty$ holds, then the function $Y(\omega) = \limsup_n X_n(\omega)$ is a random variable. If for every $\omega \in \Omega$, the condition $\inf_n X_n(\omega) > -\infty$ holds, then the function $Z(\omega) = \liminf_n X_n(\omega)$ is a random variable.*

Proof First, remember that

$$\limsup_n a_n = \inf_k \sup_{n \geq k} a_n, \text{ and}$$

$$\liminf_n a_n = \sup_k \inf_{n \geq k} a_n.$$

Thus, if $X_n(\omega)$ satisfies the assumptions, then

$$Y(\omega) = \limsup_n X_n(\omega) = \inf_k \sup_{n \geq k} X_n(\omega);$$
$$Z(\omega) = \liminf_n X_n(\omega) = \sup_k \inf_{n \geq k} X_n(\omega).$$

Hence, by Theorem 3.5 we get: $Y(\omega)$ and $Z(\omega)$ are random variables. □

Corollary 3.7 *Let X_n, $n \in \mathbb{N}$, be a sequence of random variables on $(\Omega, \mathcal{F}, \mathbf{P})$. If, for every $\omega \in \Omega$, there exists a finite limit $W(\omega) = \lim_n X_n(\omega)$, then the function $W(\omega)$ is a random variable.*

Proof It suffices to recall that a_n converges if and only if

$$\limsup_{n \to \infty} a_n = \liminf_{n \to \infty} a_n.$$

Then,

$$\lim_{n \to \infty} a_n = \limsup_{n \to \infty} a_n.$$

Thus, the result follows easily from Corollary 3.6. □

3.1.1 Exercises

145. Show that the constant function $X(\omega) \equiv c$ is a random variable on any probability space $(\Omega, \mathcal{F}, \mathbf{P})$.
146. Let $A, B \in \mathcal{F}$ be random events $(\Omega, \mathcal{F}, \mathbf{P})$, let $\mathbf{1}_A(\omega) = 1$ if $\omega \in A$, and $\mathbf{1}_A(\omega) = 0$ if $\omega \notin A$. Show that

$$\forall \omega \in \Omega \quad \mathbf{1}_{A \triangle B}(\omega) = \left(\mathbf{1}_A(\omega) - \mathbf{1}_B(\omega)\right)^2.$$

147. Let $(\Omega, \mathcal{F}, \mathbf{P})$ be a probability space and let $A \in \mathcal{F}$. Prove that the function $X(\omega) = \mathbf{1}_A(\omega)$ is a random variable.

148. Let $\Omega = [0, 1]$, $\mathcal{F} = \{\emptyset, \Omega, [0, 1/2), [1/2, 1]\}$, $\mathbf{P}$ any probability measure and let $[a]$ denote the integer part of a number a. Which of the following functions are random variables?

  (a) $X(\omega) = [\omega + 3/4]$;
  (b) $Y(\omega) = [\omega + 1/2] + 2$;
  (c) $Z(\omega) = \begin{cases} 0 & \text{if } \omega < 1/2; \\ 1 & \text{if } \omega \geq 1/2; \end{cases}$
  (d) $T(\omega) = \begin{cases} 0 & \text{if } \omega < 1/2; \\ 1 & \text{if } \omega = 1/2; \\ 2 & \text{if } \omega > 1/2. \end{cases}$

149. Show that if $X(\omega)$ and $Y(\omega)$ are random variables on $(\Omega, \mathcal{F}, \mathbf{P})$, and $Y(\omega) \neq 0$ for each $\omega \in \Omega$, then the function $X(\omega)/Y(\omega)$ is a random variable.

150. Assume that $\mathcal{A}$ is an atomic σ-field on Ω, i.e., there exists a countable or finite sequence $A_1, A_2, A_3, \ldots$ of subsets of Ω such that $\bigcup_i A_i = \Omega$, $A_i \cap A_j = \emptyset$ for $i \neq j$ and

$$\mathcal{A} = \sigma\{A_1, A_2, A_3, \ldots\}.$$

Prove that every random variable X on the probability space $(\Omega, \mathcal{A}, \mathbf{P})$ is constant on the sets A_i and that each function $X(\omega) = \sum_{i=1}^{\infty} c_i \mathbf{1}_{A_i}(\omega)$ is a random variable on $(\Omega, \mathcal{A}, \mathbf{P})$.

151. Does the fact that $|X|$ is a random variable on $(\Omega, \mathcal{F}, \mathbf{P})$ imply that X is also a random variable?

152. Show that if X and Y are random variables on $(\Omega, \mathcal{F}, \mathbf{P})$, then

$$\{\omega: X(\omega) = Y(\omega)\} \in \mathcal{F}.$$

153. Assume that the function $f : \mathbb{R} \to \mathbb{R}$ is measurable with respect to the Borel σ-field $\mathcal{B}$, i.e., $f^{-1}(B) \in \mathcal{B}$ for every $B \in \mathcal{B}$. Prove that if $X(\omega)$ is a random variable on $(\Omega, \mathcal{F}, \mathbf{P})$, then also $Y(\omega) = f(X(\omega))$ is a random variable.
Hint. Use Theorem 3.3

3.2 Distributions and Cumulative Distribution Functions

Definition 3.8 The probability distribution of a random variable X defined on a space $(\Omega, \mathcal{F}, \mathbf{P})$ is called the *probability measure* $\mathbf{P}_X$ on $(\mathbb{R}, \mathcal{B})$ and is defined by the formula

$$\mathbf{P}_X(B) \stackrel{\text{def}}{=} \mathbf{P}\left(X^{-1}(B)\right) = \mathbf{P}\{\omega : X(\omega) \in B\}, \quad B \in \mathcal{B}.$$

Sometimes, instead of $\mathbf{P}_X$, we will use the notation $\mathcal{L}(X)$.

3.2 Distributions and Cumulative Distribution Functions

Remark 3.9 It is easy to verify that any probability measure $\mathbf{Q}$ defined on the space $(\mathbb{R}, \mathcal{B})$ is the distribution of some random variable. Indeed, just take $\Omega = \mathbb{R}, \mathcal{F} = \mathcal{B}$, $\mathbf{P} = \mathbf{Q}$ and define $X(\omega) = \omega$. Then, for every $B \in \mathcal{B}$, we get:

$$\mathbf{P}_X(B) = \mathbf{P}\{\omega \in \Omega : X(\omega) \in B\} = \mathbf{Q}\{\omega \in \mathbb{R} : \omega \in B\} = \mathbf{Q}(B).$$

Remark 3.10 If for random variables X and Y defined on the same probability space $(\Omega, \mathcal{F}, \mathbf{P})$ the following condition holds:

$$\mathbf{P}\{\omega : X(\omega) = Y(\omega)\} = 1,$$

then we say that $X = Y$ *almost everywhere* with respect to the probability measure $\mathbf{P}$ (notation $X = Y$ a.e.). Two variables equal almost everywhere have identical probability distributions.

Remark 3.11 Note that a distribution does not uniquely determine a random variable, even in the sense of equality almost everywhere. This means that different random variables can have the same distributions. To see this, consider the roll of a single die and two random variables: X takes the value of one if there is an even number of pips, or zero if there is an odd number of pips; the variable Y is defined by the formula $Y(\omega) = 1 - X(\omega)$. For any Borel set B, we then have:

$$\mathbf{P}_X(B) = \mathbf{P}_Y(B) = \frac{1}{2}\mathbf{1}_B(1) + \frac{1}{2}\mathbf{1}_B(0),$$

which means that the variables X and Y have the same distribution. On the other hand, $\mathbf{P}\{\omega : X(\omega) \neq Y(\omega)\} = 1$.

Example 3.12 Suppose we bought A shares on the stock exchange for 1000 EUR. Following the previous price quotations of these stocks, we established that with a probability of $p = 0.6$ within a month, the value of these stocks will increase by ten percentage points, and with probability $1 - p = 0.4$, their value will decrease by ten percentage points. We want to describe the distribution of the random variable X determining the value of our shares three months from the date of purchase.

Let $r = 1.1$, $s = 0.9$ and $n = 1000$. If every month there is an increase, then, of course, $X = nr^3$, and the probability of this event is p^3. If there is a decline every month, then $X = ns^3$, and the probability of this event is $(1 - p)^3$. It may also happen that over the three months, the value of shares increases twice, and falls once. Then, $x = nr^2s$ with probability $3p^2(1 - p)$. For a double decrease, we get $X = nrs^2$ with a probability of $3p(1 - p)^2$. The distribution of the random variable X can be written in the form of a table:

k	1331	1089	891	729
$\mathbf{P}(X = k)$	0.216	0.432	0.288	0.064

Instead of the above notation, we can use a convex combination of δ_x measures that assign the measure 1 to the one-point set $\{x\}$, $x \in \mathbb{R}$,

$$\mathbf{P}_X = 0.216 \cdot \delta_{nr^3} + 0.432 \cdot \delta_{nr^2s} + 0.288 \cdot \delta_{nrs^2} + 0.064 \cdot \delta_{ns^3}.$$

We can also use the obvious relationships of the obtained distribution with the already known Bernoulli distribution and write

$$\mathbf{P}\left\{X = nr^k s^{3-k}\right\} = \binom{3}{k} p^k (1-p)^k, \quad k = 0, 1, 2, 3.$$

We will describe the distribution of discrete random variables in a similar way.

Definition 3.13 A random variable X is *simple* if the set of its values is finite. A random variable X is *discrete* if the set of its values is countable.

The number of heads thrown in n coin tosses, or the number of spades received in a bridge hand are random, simple and discrete variables. On the other hand, the number of times you toss a coin to get heads for the first time is a discrete random variable, but not a simple one because the set of values for this variable is countable and infinite.

One of the most convenient methods of describing the distribution of a random variable is to use its (cumulative) distribution function.

Definition 3.14 The *(cumulative) distribution function* of a random variable X living on the space $(\Omega, \mathcal{F}, \mathbf{P})$ is the function $F : \mathbb{R} \to [0, 1]$ defined by

$$F(t) = \mathbf{P}\{\omega : X(\omega) < t\}.$$

In many textbooks, a weak inequality appears in the definition of the cumulative distribution function F. Both approaches are acceptable, but for the purposes of this book, we chose the sharp inequality. Sometimes it is more convenient to say that F is the distribution of the probabilistic measure $\mathbf{P}_X$ if $\mathbf{P}_X$ is the distribution of the random variable X. Then, we write

$$F(t) = \mathbf{P}_X((-\infty, t)).$$

Example 3.15 Consider the number of heads thrown in a toss of two symmetric coins. The space of elementary events Ω can be identified with the set $\{0, 1, 2\}$, and the probability $\mathbf{P}$ is given by the formula: $\mathbf{P}(A) = \frac{1}{4}\delta_0(A) + \frac{1}{4}\delta_2(A) + \frac{1}{2}\delta_1(A)$ for any $A \subset \Omega$. Note that $\Omega \subset \mathbb{R}$, so we can also consider $\mathbf{P}$ as a probability measure on $\mathbb{R}$. If we define $X : \mathbb{R} \mapsto \mathbb{R}$ by $X(\omega) = \omega$, then we can see that $\mathbf{P}_X = \mathbf{P}$. The

3.2 Distributions and Cumulative Distribution Functions

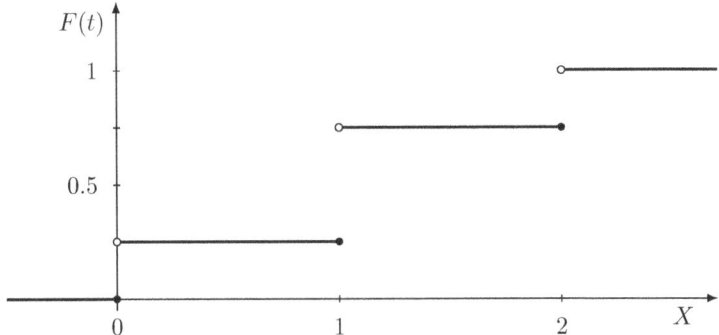

Fig. 3.1 The graph of $F(t)$

cumulative distribution of the measure **P** is equal to the distribution function of the random variable X and is given by:

$$F(t) = \frac{1}{4}\mathbf{1}_{(0,\infty)}(t) + \frac{1}{2}\mathbf{1}_{(1,\infty)}(t) + \frac{1}{4}\mathbf{1}_{(2,\infty)}(t) = \begin{cases} 0, & t \leq 0; \\ 0.25, & 0 < t \leq 1; \\ 0.75, & 1 < t \leq 2; \\ 1, & t > 2. \end{cases}$$

Figure 3.1 shows the graph of the function $F(t)$.

Theorem 3.16 *Let F be the cumulative distribution function of a random variable X on $(\Omega, \mathcal{F}, \mathbf{P})$. Then,*

(1) *F is nondecreasing;*
(2) *$\lim_{t \to -\infty} F(t) = 0$, $\lim_{t \to \infty} F(t) = 1$;*
(3) *F is left-continuous, i.e., $F(t) = \lim_{s \nearrow t} F(s)$ for every $t \in \mathbb{R}$.*

Proof Let $s < t$. Then, $(-\infty, t) = (-\infty, s) \cup [s, t)$. Thus, since the probability measure $\mathbf{P}_X$ is additive and non-negative, we get

$$F(t) = \mathbf{P}_X((-\infty, s)) + \mathbf{P}_X([s, t)) \geq \mathbf{P}_X((-\infty, s)) = F(s),$$

which proves property (1). In order to prove (2), consider a decreasing sequence of real numbers $x_1 > x_2 > \cdots$ such that $\lim_{n \to \infty} x_n = -\infty$. Then, $\{(-\infty, x_n)\}$ is a decreasing sequence of sets and $\bigcap_{n=1}^{\infty}(-\infty, x_n) = \emptyset$. Now, by the Continuity Theorem for probability measures we conclude that

$$0 = \mathbf{P}_X(\emptyset) = \mathbf{P}_X\left(\bigcap_{n=1}^{\infty}(-\infty, x_n)\right) = \lim_{n \to \infty} \mathbf{P}_X((-\infty, x_n)) = \lim_{n \to \infty} F(x_n).$$

In a similar way, we prove the second equality from condition (2). For any increasing sequence of real numbers $x_1 < x_2 < \cdots$ such that $x_n \to \infty$, the sequence of sets $\{(-\infty, x_n)\}$ is increasing and $\bigcup_{n=1}^{\infty}(-\infty, x_n) = \mathbb{R}$, hence

$$1 = \mathbf{P}_X(\mathbb{R}) = \mathbf{P}_X\left(\bigcup_{n=1}^{\infty}(-\infty, x_n)\right) = \lim_{n \to \infty} \mathbf{P}_X((-\infty, x_n)) = \lim_{n \to \infty} F(x_n).$$

To prove (3), take $t \in \mathbb{R}$ and any increasing sequence of real numbers $x_n \nearrow t$. The sequence of sets $\{(-\infty, x_n)\}$ is increasing and such that $\bigcup_{n=1}^{\infty}(-\infty, x_n) = (-\infty, t)$, hence, using the Continuity Theorem again, we get

$$F(t) = \mathbf{P}_X\left(\bigcup_{n=1}^{\infty}(-\infty, x_n)\right) = \lim_{n \to \infty} \mathbf{P}_X((-\infty, x_n)) = \lim_{n \to \infty} F(x_n),$$

from which the left-continuity of the function F follows. □

Theorem 3.17 *If a function* $F : \mathbb{R} \to \mathbb{R}$ *satisfies conditions* (1), (2) *and* (3) *of Theorem 3.16, then it is a cumulative distribution function of some random variable.*

Proof We present here a non-constructive proof of this theorem. However, since every mathematician should be able to construct a measure directly from its cumulative distribution function, we also present the full constructive proof in Chap. 8.

Let F be a function satisfying conditions (1), (2), and (3). For the space Ω, we will take the interval $(0, 1)$, and let the probability measure $\mathbf{P}$ be the Lebesgue measure on Ω. We define:

$$X(\omega) = \sup\{t : F(t) \leq \omega\}.$$

It is easy to see that X is a random variable on the space $(\Omega, \mathcal{F}, \mathbf{P})$, where $\mathcal{F}$ is the σ-field of Borel subsets of the set $(0, 1)$. Furthermore, X is a non-decreasing function and $X(\omega) < r$ if and only if $F(r) > \omega$. Hence,

$$\mathbf{P}\{\omega : X(\omega) < r\} = \mathbf{P}\{\omega : \sup\{t : F(t) \leq \omega\} < r\} = \mathbf{P}\{\omega : F(r) > \omega\} = F(r),$$

which completes the proof of the theorem. □

Let us also consider the discontinuity of the distribution function (see Fig. 3.2). We know that each distribution function F of any random variable X is a left-continuous and non-decreasing function. Therefore, if $t_0 \in \mathbb{R}$ is a discontinuity point, then

$$F(t_0) = \lim_{t \nearrow t_0} F(t) < \lim_{t \searrow t_0} F(t).$$

3.2 Distributions and Cumulative Distribution Functions

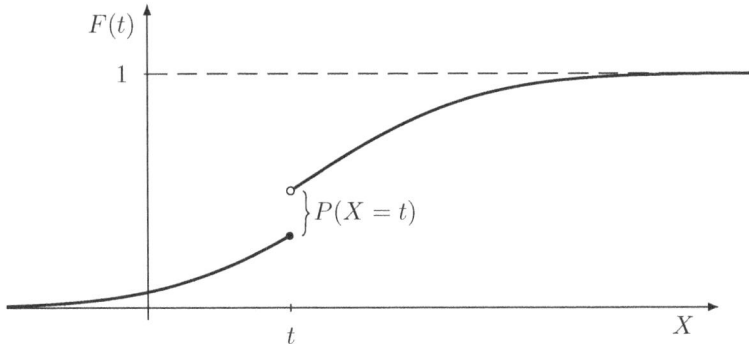

Fig. 3.2 Graph indicating the discontinuity of the distribution function $F(t)$

The right-hand limit exists due to the monotonicity of the function F. For any sequence $t_n \searrow t_0$, the sequence of sets $[t_0, t_n)$ is decreasing, hence

$$\mathbf{P}\{X = t_0\} = \mathbf{P}\left\{X \in \bigcap_{n=1}^{\infty} [t_0, t_n)\right\} = \lim_{n \to \infty} \mathbf{P}\{X \in [t_0, t_n)\}$$

$$= \lim_{n \to \infty} \left[F(t_n) - F(t_0)\right] = \lim_{n \to \infty} F(t_n) - F(t_0).$$

In this way, we have proved that if t_0 is a discontinuity point of the cumulative distribution function F, then the size of this jump is equal to $\mathbf{P}\{X = t_0\}$. This also means that the probability measure $\mathbf{P}_X$ has at the point t_0 an atom with weight equal to the cumulative distribution function's jump. If the cumulative function of X is continuous, then $\mathbf{P}_X$ has no atoms.

3.2.1 Exercises

154. Let $\Omega = \{0, 1, 2, 3\}$, and $\mathbf{P}\{k\} = \frac{1}{4}$ for $k = 0, 1, 2, 3$. We define random variables: $X(\omega) = \sin\frac{\pi\omega}{2}$ and $Y(\omega) = \cos\frac{\pi\omega}{2}$. Find the distributions and the cumulative distribution functions for X and Y. Calculate $\mathbf{P}\{\omega : X(\omega) = Y(\omega)\}$.

155. Determine the set of all triples $(a, b, c) \in \mathbb{R}^3$ for which the following function is a distribution function of some random variable:

(a) $F_1(t) = \begin{cases} at^2, & t \leq 0; \\ bt + c, & 0 < t \leq 2; \\ 1, & t > 2; \end{cases}$ (b) $F_2(t) = \begin{cases} 0, & t < 0; \\ a\sin t + b, & 0 \leq t \leq \pi/2; \\ 1, & t > \pi/2. \end{cases}$

156. Let $\Omega = [0, 3]$ and let **P** be the normalized Lebesgue measure on Ω. Find the distribution functions of the following random variables:

(a) $X(\omega) = \begin{cases} 2\omega + 1, & 0 \leqslant \omega \leqslant 1; \\ -\omega^2 + 2, & 1 < \omega < 2; \\ 3, & 2 \leqslant \omega \leqslant 3; \end{cases}$ (b) $Y(\omega) = \begin{cases} -\omega + 1, & 0 \leqslant \omega < 1; \\ \omega^2 - 1, & 1 \leqslant \omega \leqslant 2; \\ 3\omega, & 2 < \omega \leqslant 3. \end{cases}$

157. Let $\Omega = [0, 1]$ and let **P** be the Lebesgue measure on Ω. Find the cumulative distribution functions of the following random variables: (a) $X(\omega) = |2\omega - 1|$; (b) $Y(\omega) = \sin^2 \omega + \cos^2 \omega$; (c) $Z(\omega) = 1$ if $\omega \in Q$; $Z(\omega) = 0$, if $\omega \in Q'$.
158. A random variable X has distribution function F. Find the distribution functions of the following random variables: (a) $Y(\omega) = aX(\omega) + b$; (b) $Z(\omega) = aX^2(\omega)$.
159. Prove that the distribution function F of the probability measure **P** can have at most countably many points of discontinuity.
160. Does there exist a distribution function whose set of discontinuity points is dense in $\mathbb{R}$?
161. Let X be a random variable with continuous distribution function F. Show that for any countable set $A \subset \mathbb{R}$, we have $\mathbf{P}\{\omega: X(\omega) \in A\} = 0$.
162. A random variable X has distribution function F. Find the cumulative distribution function of the random variable $Y = \frac{1}{2}(X + |X|)$.
163. Functions F and G are distribution functions of some random variables. Give the necessary and sufficient conditions for the function $H(x) := F(G(x))$ to be a distribution function as well.

3.3 Review of Discrete Distributions

In this section, we will discuss the more important discrete distributions. We have already seen some of them. The ones that appear for the first time will be described in more detail. Recall that $\mathbf{1}_A(x)$ stands for the function that takes the value 1 if $x \in A$ and zero otherwise.

3.3.1 Single Point Distributions

A random variable X has a *single point distribution* (or *one-point distribution*) if there is an $a \in \mathbb{R}$ such that

$$\mathbf{P}\{X = a\} = 1.$$

3.3 Review of Discrete Distributions

Of course, such a random variable is constant almost everywhere with respect to the probability measure **P**. This does not mean, however, that it has to be a constant function. For example, if $\Omega = [0, 1]$, **P** is Lebesgue measure and $X(\omega) = a$ for irrational numbers ω and $X(\omega) = a - 1$ for ω rational, then we can see that $\mathbf{P}\{X = a\} = 1$, but $X(\omega) \neq a$ on a quite large set.

A one-point distribution can be written in the following way: $\mathbf{P}_X = \delta_a$, and its distribution function has the form:

$$F(t) = \mathbf{1}_{(a,\infty)}(t).$$

3.3.2 Two-Point Distributions

A random variable X has a *two-point distribution* if it can only take two values, a and b, where

$$\mathbf{P}\{X = a\} = p = 1 - \mathbf{P}\{X = b\}.$$

We can also describe it in the following way:

$$\mathbf{P}_X = p\delta_a + (1-p)\delta_b, \qquad F_X(t) = p\mathbf{1}_{(a,\infty)}(t) + (1-p)\mathbf{1}_{(b,\infty)}(t).$$

If the random variable X describes a Bernoulli trial, success is usually assigned the value of $a = 1$, and failure the value of $b = 0$.

3.3.3 Binomial Distributions or Bernoulli Distributions

A variable X has a *binomial distribution* with parameters $n \in \mathbb{N}$ and $p \in (0, 1)$ if

$$\mathbf{P}\{X = k\} = \binom{n}{k} p^k (1-p)^{n-k}, \quad k = 0, 1, \ldots, n$$

or

$$P_X = \sum_{k=0}^{n} \binom{n}{k} p^k (1-p)^{n-k} \delta_k.$$

This random variable (which was described in detail in Sect. 2.9) counts the number of successes in n Bernoulli trials.

3.3.4 Multinomial Distributions

The *multinomial distribution* is a generalization of the binomial distribution. In a Bernoulli trial, the outcome can be either success or failure. Here, we assume that the result of a trial can be one of k possible results, where the i-th result appears with probability p_i, $i = 1, \ldots, k$, and this probability cannot change from trial to trial. If we make n independent trials and denote the number of results of the i-th type by X_i, then we can write

$$\mathbf{P}\{X_1 = n_1, \ldots, X_k = n_k\} = \frac{n!}{n_1! \ldots n_k!} p_1^{n_1} \ldots p_k^{n_k},$$

where $p_i \in (0, 1)$, $p_1 + \cdots + p_k = 1$, $n_1 + \cdots + n_k = n$. For example, if we know that there are 20% eels, 25% roaches and 55% breams in a pond, and Jan caught 10 fish, the vector (*number of eels, number of roaches, number of breams*) will have a multinomial distribution with parameters $n = 10$; $k = 3$; $p_1 = 0.2$; $p_2 = 0.25$ and $p_3 = 0.55$.

3.3.5 Poisson Distributions

A random variable X has a *Poisson distribution* with parameter $\lambda > 0$ if

$$\mathbf{P}\{X = k\} = \frac{\lambda^k}{k!} \cdot e^{-\lambda}, \quad k = 0, 1, 2, \ldots \quad \text{or} \quad \mathbf{P}_X = \sum_{k=0}^{\infty} \frac{\lambda^k}{k!} \cdot e^{-\lambda} \delta_k.$$

It turns out that the Poisson distribution with parameter $\lambda = \lambda_0 t$ describes well the number of calls received at a telephone exchange over time t, or the number of radioactive particles registered during time t. With such an interpretation of the variable X, the parameter λ_0 is called the intensity of the distribution.

3.3.6 Geometric Distributions

A random variable X has a *geometric distribution* with parameter $p \in (0, 1)$ if

$$\mathbf{P}\{X = k\} = pq^{k-1}, \quad k = 1, 2, \ldots \quad \text{or} \quad \mathbf{P}_X = \sum_{k=1}^{\infty} pq^{k-1} \delta_k.$$

This is the distribution of the waiting time for the first success in an infinite sequence of Bernoulli trials. Note that regardless of the value of $p \in (0, 1)$, the most likely value for X is $k = 1$. There is another version of the geometric distribution in which

3.3 Review of Discrete Distributions

a random variable counts the number of failures before the first success, i.e.,

$$\mathbf{P}\{Y = k\} = pq^k, \quad k = 0, 1, 2, \ldots \quad \text{or} \quad \mathbf{P}_X = \sum_{k=0}^{\infty} pq^k \, \delta_k.$$

3.3.7 Pascal (Negative Binomial) Distributions

A random variable X has a *Pascal distribution* with parameters $r \in \mathbb{N}$ and $p \in (0, 1)$ if

$$\mathbf{P}\{X = k\} = \binom{k-1}{r-1} p^r (1-p)^{k-r}, \quad k = r, r+1, r+2, \ldots$$

or

$$\mathbf{P}_X = \sum_{k=r}^{\infty} \binom{k-1}{r-1} p^r (1-p)^{k-r} \, \delta_k.$$

The Pascal distribution describes the waiting time for the r-th success in a sequence of Bernoulli trials. The geometric distribution is a special case of the Pascal distribution when $r = 1$.

The generalized Pascal distribution does not require the assumption that r is a natural number; however, we then replace the factorial in the above formula by using the function Γ, which will be discussed in Sect. 3.4.3. We then say that the variable X has a *generalized Pascal distribution* with parameters $r > 0$ and $p \in (0, 1)$ if

$$\mathbf{P}\{X = k\} = \frac{\Gamma(k)}{\Gamma(r)\Gamma(k-r)} \cdot p^r (1-p)^{k-r}, \quad k = r, r+1, r+2, \ldots$$

3.3.8 Hypergeometric Distributions

A random variable X has a *hypergeometric distribution* with parameters $M, N, n \in \mathbb{N}, n \leqslant N, n \leqslant M$ if

$$\mathbf{P}\{X = k\} = \frac{\binom{N}{k}\binom{M}{n-k}}{\binom{N+M}{n}}, \quad k = 0, 1, 2, \ldots, n.$$

We have seen such a distribution before, for example, when calculating the probability of getting four spades in a bridge hand. We then have: $N = 13$; $M = 39$; $k = 4$ and $n = 13$.

3.3.9 Exercises

164. Find the cumulative distribution functions for the following discrete distributions: (a) two-point; (b) geometric; (c) Bernoulli.
165. Find the most probable value for the Pascal distribution with parameters r and p.
166. The random variable X describes the number of aces we can get in a bridge hand (13 cards). Find the distribution of X.
167. A random variable X takes values in $\mathbb{N}_0 = \{0, 1, \ldots\}$. Show that the following conditions are equivalent:

    (1) X has a geometric distribution: $\mathbf{P}\{X = k\} = pq^k, k = 0, 1, \ldots$;
    (2) $\mathbf{P}\{X = n + k | X \geqslant k\} = \mathbf{P}\{X = n\}$.

168. Tim wants to buy a computer game, but he is 3 euros short. Every evening he persuades his mother to give him one euro, and with probability 0.5 he gets it. What is the probability that he will buy the game later this week if it is Tuesday and the store is only closed on Sunday?
169. Paul has 10 coins and two money boxes. He scatters the coins on a table and puts all those that have fallen tails up into the first piggy bank and the others into the second. Find the distribution of X, the number of coins in the first money box.
170. Jane believes that a success in a dice roll is when she gets a number divisible by 3, and in a coin toss, when she gets heads. She chooses randomly whether she will roll a die or toss a coin, and the probability that she chooses a coin is 0.5. What is the distribution of the random variable X describing the number of experiments until she has been successful three times?

3.4 Continuous Type Distributions

Definition 3.18 A random variable $X : \Omega \to \mathbb{R}$ is of *continuous type* or has a *continuous type distribution* if there exists a measurable and integrable function $f : \mathbb{R} \to [0, \infty)$ such that for every Borel set $B \subset \mathbb{R}$

$$\mathbf{P}_X(B) = \mathbf{P}\{X(\omega) \in B\} = \int_B f(x)\,dx.$$

The function f is called the *density* of the random variable X or the density of the probability distribution $\mathbf{P}_X$ on the real line.

3.4 Continuous Type Distributions

It is easy to see that $\int_{-\infty}^{\infty} f(x)\,\mathrm{d}x = 1$. Moreover, any non-negative integrable function f on the line $\mathbb{R}$ satisfying the condition $\int_{-\infty}^{\infty} f(x)\,\mathrm{d}x = 1$ is the density of some random variable X (it is the density of some probability distribution $\mathbf{P}_X$). It suffices to define for any Borel set B:

$$\mathbf{P}_X(B) = \int_B f(x)\,\mathrm{d}x.$$

Often, instead of saying that a random variable is of continuous type, we say that it is *absolutely continuous*. This is related to the absolute continuity of the distribution of this variable with respect to Lebesgue measure. This concept will not be discussed in more detail until Chap. 8. Here, we will limit ourselves only to the following statement:

Theorem 3.19 *If a continuous function F is a distribution function of some random variable such that the function $f(x) = F'(x)$ exists on the entire line $\mathbb{R}$ except for a Lebesgue measure zero set, and*

$$\int_{-\infty}^{\infty} f(x)\,\mathrm{d}x = 1,$$

then f is the density function for the distribution function F.

Proof It suffices to note that for any a, b, $a < b$, by virtue of Fatou's Lemma, the following inequality holds:

$$\begin{aligned}
\int_a^b f(x)\,\mathrm{d}x &= \int_a^b \liminf_{h \searrow 0} \frac{F(x+h) - F(x)}{h}\,\mathrm{d}x \\
&\leqslant \liminf_{h \searrow 0} \int_a^b \frac{F(x+h) - F(x)}{h}\,\mathrm{d}x \\
&= \liminf_{h \searrow 0} \frac{G(b+h) - G(b) - G(a+h) + G(a)}{h} \\
&\leqslant \liminf_{h \searrow 0} [F(b + \theta_1 h) - F(a + \theta_2 h)] = F(b) - F(a),
\end{aligned}$$

where G is a primitive function for the continuous function F, and the existence of $\theta_1, \theta_2 \in (0, 1)$ is due to the mean value theorem. From the assumptions of our theorem and the obtained inequality, we get:

$$\int_{-\infty}^{t} f(x)\,\mathrm{d}x \leqslant F(t) - F(-\infty) = F(t)$$

and

$$\int_{-\infty}^{t} f(x)\,\mathrm{d}x = 1 - \int_{t}^{\infty} f(x)\,\mathrm{d}x \geqslant 1 - (F(\infty) - F(t)) = F(t).$$

□

We now discuss the most important distributions of continuous type.

3.4.1 Uniform Distributions on an Interval

A *uniform distribution* is a distribution with parameters a, b, $a < b$, described by the density:

$$f(x) = \begin{cases} \frac{1}{b-a} & \text{if } a \leqslant x \leqslant b; \\ 0 & \text{if } x < a \text{ or } x > b. \end{cases}$$

This distribution is identical with the measure we call the geometric probability on the interval $[a, b]$, or more generally, with the probability equal to the normalized Lebesgue measure on the interval $[a, b]$, because

$$\mathbf{P}_X([c, d]) = \int_c^d f(x)\,\mathrm{d}x = \frac{d-c}{b-a} \quad \text{for every } [c, d] \subset [a, b].$$

3.4.2 Exponential Distributions

A random variable X has an *exponential distribution* with parameter $\lambda > 0$ if it has a density function given by

$$f(x) = \begin{cases} \lambda e^{-\lambda x} & \text{if } x > 0; \\ 0 & \text{if } x \leqslant 0. \end{cases}$$

This variable takes only positive values (with probability one). Equivalently, we can say that the positive half-line is the support of this distribution.

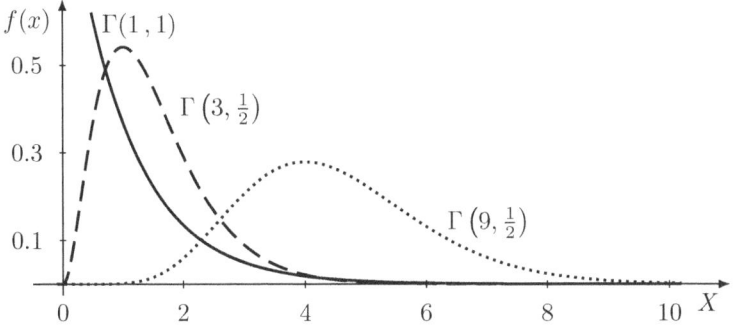

Fig. 3.3 The gamma distribution for various parameters

3.4.3 Gamma Distributions

The *gamma distribution* with parameters $p, b > 0$ has the density function:

$$f(x) = \begin{cases} \frac{b^p}{\Gamma(p)} x^{p-1} e^{-bx} & \text{if } x > 0; \\ 0 & \text{if } x \leqslant 0. \end{cases}$$

We use the notation $\Gamma(p, b)$ in this case. Note that the $\Gamma(1, b)$ distribution is the exponential distribution with parameter b. Figure 3.3 shows some examples of gamma densities for various parameters.

The function Γ appearing in the gamma density function is defined by

$$\Gamma(p) \stackrel{\text{def}}{=} \int_0^\infty x^{p-1} e^{-x} \, dx, \quad p > 0.$$

It is easy to see that $\Gamma(1) = \Gamma(2) = 1$. Moreover,

$$\Gamma(p+1) = p \, \Gamma(p).$$

This formula describes the properties of the Γ function sufficiently for our purposes. Its proof lies in the simple application of integration by parts. By mathematical induction, we get $\Gamma(n+1) = n!$ for every $n \in \mathbb{N}$.

Since for $p \notin \mathbb{N}$ integration of the function $f(x)$ is not easy, it is worth remembering that

$$\int_0^\infty x^{p-1} e^{-bx} \, dx = \frac{\Gamma(p)}{b^p},$$

as $f(x)$ is a probability density function, and $\int_\mathbb{R} f(x)\,dx = 1$. We will make use of this fact many times when calculating the parameters of these probability distributions.

3.4.4 Beta Distributions

The *beta distribution* with parameters $p, q > 0$ lives on the interval $[0, 1]$, is denoted by $\mathcal{B}e(p, q)$ and has a density of the form:

$$f(x) = \begin{cases} \frac{\Gamma(p+q)}{\Gamma(p)\Gamma(q)} x^{p-1}(1-x)^{q-1} & \text{if } 0 < x < 1; \\ 0 & \text{if } x \leqslant 0 \text{ or } x \geqslant 1. \end{cases}$$

Figure 3.4 shows some examples of the densities of such distributions.

The name of the distribution comes from the binary beta function $\mathcal{B}(p, q)$, which generalizes the symbol $\binom{n}{k}$ and is defined by the formula:

$$\mathcal{B}(p, q) = \frac{\Gamma(p)\Gamma(q)}{\Gamma(p+q)}, \quad p, q > 0.$$

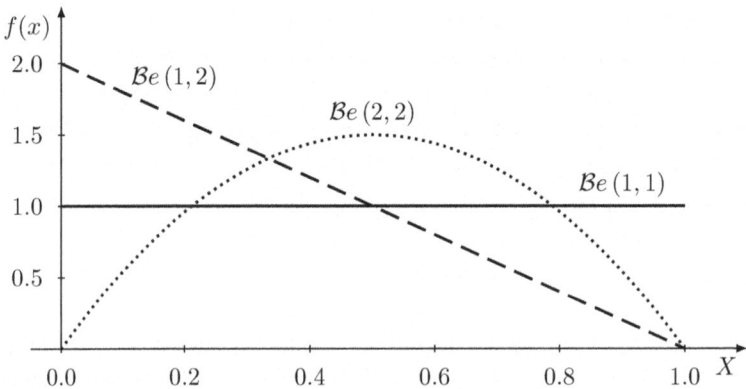

Fig. 3.4 The beta distribution for various parameters

3.4.5 Cauchy Distributions

A random variable X has a *Cauchy distribution* with parameters $a > 0$ and $m \in \mathbb{R}$ if its density has the form:

$$f(x) = \frac{a}{\pi(a^2 + (x-m)^2)}.$$

Some examples are given in Fig. 3.5.

The constant a in this density function is called the *scale parameter*, and the parameter m specifies the point with respect to which the distribution is symmetric. It is easy to see that $\mathbf{P}\{X > m\} = \mathbf{P}\{X < m\} = \frac{1}{2}$.

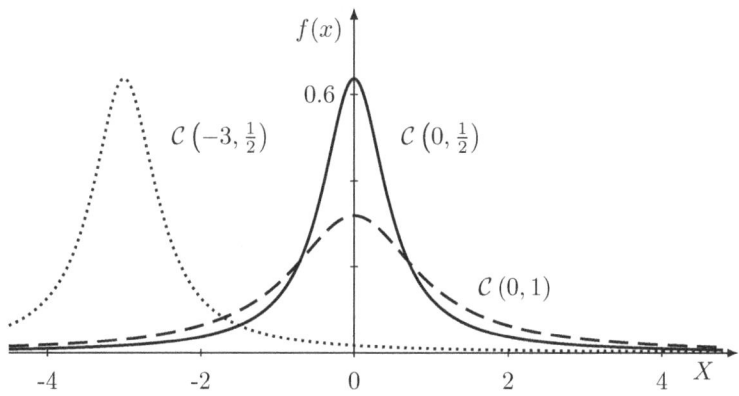

Fig. 3.5 The Cauchy distribution for various parameters

3.4.6 Gaussian Distributions

A random variable X has a *normal*, or *Gaussian distribution*, with parameters $m \in \mathbb{R}$ and $\sigma > 0$ if the function

$$f(x) = \frac{1}{\sqrt{2\pi}\sigma} \exp\left\{-\frac{(x-m)^2}{2\sigma^2}\right\}$$

is the density function of this distribution. We denote such a distribution symbolically by $\mathcal{N}(m, \sigma)$.

Figure 3.6 shows a few example graphs of the function $f(x)$ for various parameters σ. The parameter m is always the midpoint of the graph.

Note that it is not easy to show that $\int_{\mathbb{R}} f(x)\,dx = 1$ because the primitive function Φ of the function f is not elementary. There is, however, a particularly

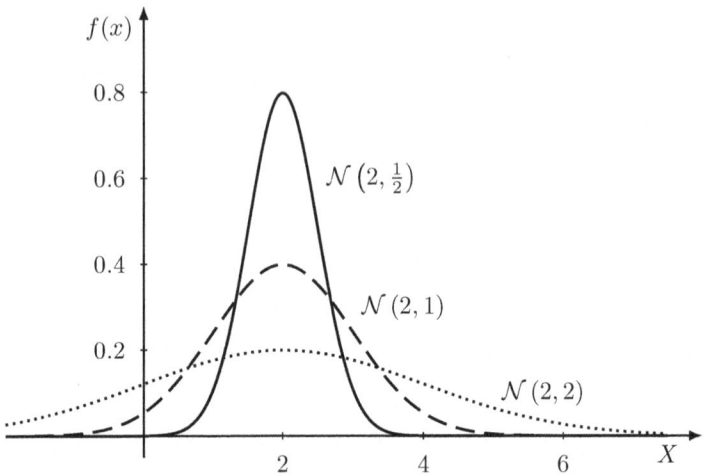

Fig. 3.6 The Gaussian distribution for various parameters

nice and simple way to compute this integral, namely:

$$\left(\int_{-\infty}^{\infty} e^{-x^2/2}\,dx\right)^2 = 4\int_0^{\infty} e^{-x^2/2}\,dx \int_0^{\infty} e^{-y^2/2}\,dy$$

$$= 4\int\cdots\int_{\mathbb{R}_+^2} e^{-(x^2+y^2)/2}\,dx\,dy = 4\int_0^{\infty}\int_0^{\pi/2} e^{-r^2/2} r\,d\varphi\,dr$$

$$= 4\frac{\pi}{2}\left[-e^{-r^2/2}\right]_0^{\infty} = 2\pi.$$

Of course, when integrating over the region $\mathbb{R}_+^2$, we switched to polar coordinates using the substitution $x = r\cos\varphi$, $y = \sin\varphi$. Finally, we have

$$\int_{-\infty}^{\infty} e^{-x^2/2}\,dx = \sqrt{2\pi}.$$

Now, to show that $\int_{\mathbb{R}} f(x)\,dx = 1$, we simply substitute $y = \frac{x-m}{\sigma}$ and use the resulting formula.

The cumulative distribution function for the distribution $\mathcal{N}(0,1)$ is given by:

$$\Phi(t) = \int_{-\infty}^{t} \frac{1}{\sqrt{2\pi}} e^{-x^2/2}\,dx.$$

3.4 Continuous Type Distributions

Since the density of the distribution $\mathcal{N}(0, 1)$ is a symmetric function, the following equality holds:

$$\Phi(-t) = 1 - \Phi(t).$$

Φ is a special function and cannot be written in the language of elementary functions. Tables of values of Φ can be found in Chap. 9.

The normal distribution plays a very important role in probability theory and mathematical statistics. It is sometimes said, somewhat informally, that if a random variable is the sum of a very large number of very small independent terms, then it has a normal distribution (see Sect. 7.2).

3.4.7 Exercises

171. A random variable X has an exponential distribution with parameter $\lambda > 0$. What is the distribution of $Y = \lambda^{-1} X$?
172. A random variable X has a gamma distribution $\Gamma(b, p)$. What is the distribution of $Y = b^{-1} X$?
173. A random variable X has a Cauchy distribution with parameters $a > 0$ and $m \in \mathbb{R}$. What is the distribution of $Y = a^{-1}(X - m)$?
174. Find the cumulative distribution functions of the following distributions: (a) the uniform distribution on the interval $[a, b]$, $a < b$; (b) the exponential distribution; (c) the symmetric Cauchy distribution.
175. A continuous type random variable X has a distribution function F and positive everywhere density function f. Find the densities of the following variables: (a) $Y(\omega) = aX(\omega) + b$; (b) $Z(\omega) = aX^2(\omega)$; (c) $W(\omega) = F(X(\omega))$.
176. A random variable X has a gamma distribution with parameters $p = 17$, $b = 77$. What is the distribution of $Y = \cos^2 X + \sin^2 X$?
177. A random variable X has a uniform distribution on the interval $[0, 1]$. Find the distribution of $Y = \lambda^{-1} \ln(X)$.
178. Prove that if a random variable X has a uniform distribution on the interval $[-\frac{\pi}{2}, \frac{\pi}{2}]$, then the variable $Y = \text{tg} X$ has a symmetric Cauchy distribution.
179. A random variable X has a Gaussian distribution $\mathcal{N}(m, \sigma)$. Show that the variable $Y = aX + b$, $a \neq 0$, also has a normal distribution and determine the parameters of this distribution.
180. A random variable X is exponentially distributed with parameter $\lambda = 4$. Find the distribution function for the random variable $Y = \sqrt{X}$. Is the variable Y absolutely continuous? If so, find its density.
181. An isosceles triangle with a vertex at the origin of the coordinate system has a side of length 1 contained in the non-negative part of the OX axis. The other side of length 1 starts at the origin and lies at a random angle α to the OX axis. Find the density of the random variable which determines the length of

the third side if the random variable α has a uniform distribution on the interval $[0, 2\pi]$

182. We choose at random a point $\omega = (x, y)$ from the square $K = [0, 1] \times [0, 1]$. If $x^2 + y^2 \leq 1$, then we define $X(\omega) = \sqrt{x^2 + y^2}$. If $x^2 + y^2 > 1$, then $X(\omega)$ is equal to the distance of the point (x, y) from the nearest side of the square K. Find the distribution of the random variable X. Is X of continuous type?

183. A random variable X has a normal distribution $N(m, \sigma)$. Find the distribution of the variable $Y = h(X)$, where

(a) $h(x) = \dfrac{x}{|x|}$, (b) $h(x) = \begin{cases} ca, & a < x; \\ cx, & -a \leq x \leq a; \\ -ca, & x < -a. \end{cases}$

3.5 A Complete Description of the Types of Random Variables

Discrete random variables, continuous type variables and their mixtures do not cover all types of random variables. In the past, more often than not, the third type of distribution, i.e. the singular distributions, were omitted in textbooks on probability calculus because they were considered rather absent in the modeling of real phenomena. Today, we know that such distributions appear in stock exchange quotes, atmospheric phenomena, environmental pollution processes and in modeling the widening of the ozone hole. Therefore, in this book, we also consider singular distributions.

Definition 3.20 A random variable X has a *singular distribution* if X takes values in an uncountable set A such that $\lambda_1(A) = 0$ (e.g. A can be the Cantor set), where λ_1 is the Lebesgue measure on $\mathbb{R}$, and

$$\mathbf{P}\{X \in A\} = 1, \qquad \mathbf{P}\{X = x\} = 0 \text{ for each } x \in A.$$

Example 3.21 In mathematics, the *Cantor function* is an example of a function that is continuous, but not absolutely continuous. It is continuous everywhere and has zero derivative almost everywhere, and yet it grows on the interval [0,1] from zero to one. It is also referred to as the *Cantor ternary function*, the *Lebesgue function*, *Lebesgue's singular function*, the *Cantor–Vitali function*, the *Devil's staircase*, the *Cantor staircase function*, and the *Cantor–Lebesgue function*. We interpret the Cantor function as the cumulative distribution function of a singular measure.

3.5 A Complete Description of the Types of Random Variables

First, we construct the Cantor function $c: [0, 1] \mapsto [0, 1]$. For $x \in [0, 1]$, we obtain $c(x)$ by the following steps:

(1) Express x in base 3.
(2) If the base 3 representation of x contains a 1, replace every digit strictly after the first 1 with 0.
(3) Replace any remaining 2s with 1s.
(4) Interpret the result as a binary number. The result is $c(x)$.

For example: the number $\frac{1}{4}$ has ternary representation $0.02020202\ldots$ There are no ones in this representation, so the second step still gives us $0.02020202\ldots$ This is rewritten as $0.01010101\ldots$, which is the binary representation of $\frac{1}{3}$, thus $c(1/4) = \frac{1}{3}$.

The number $\frac{1}{5}$ has ternary representation $0.01210121\ldots$ The digits after the first 1 are replaced with zeros to produce $0.01000000\ldots$ This is not rewritten since it has no 2s. It is the binary representation of $\frac{1}{4}$, thus $c(1/5) = \frac{1}{4}$.

Equivalently, if C is the Cantor set on $[0, 1]$, then the Cantor function (see Fig. 3.7) $c: [0, 1] \mapsto [0, 1]$ can be defined as:

$$c(x) = \begin{cases} \sum_{n=1}^{\infty} \frac{a_n}{2^n} & \text{if } x = \sum_{n=1}^{\infty} \frac{a_n}{3^n} \text{ for } a_n \in \{0, 1\}; \\ \sup\{c(y): y \leq x, y \in C\} & \text{if } x \in [0, 1] \setminus C. \end{cases}$$

This formula is well-defined, since every member of the Cantor set has a unique base 3 representation that only contains the digits 0 or 2. Since $c(0) = 0$ and $c(1) = 1$ and c is monotonic on C, it is also clear that $0 \leq c(x) \leq 1$ for all $x \in [0, 1] \setminus C$. Now, we can define the corresponding cumulative distribution function:

$$F_c(t) = c(t)\mathbf{1}_{[0,1]}(t) + \mathbf{1}_{(1,\infty)}(t).$$

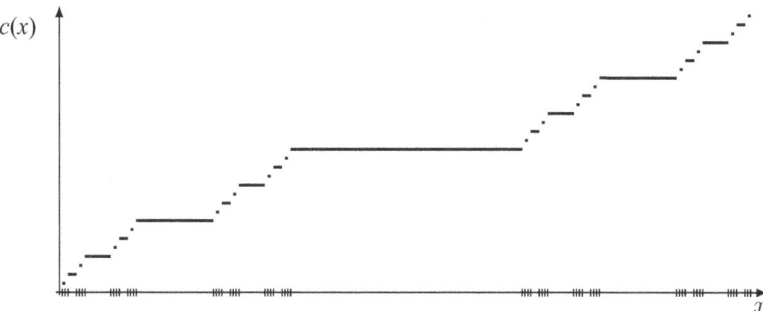

Fig. 3.7 The Cantor function

Most important for us is the following decomposition theorem:

Theorem 3.22 (Lebesgue Decomposition Theorem) *For every probability measure* $\mathbf{P}$ *there exists an atomic probability measure* $\mathbf{P}_{at}$, *a probability measure of continuous type* $\mathbf{P}_{ct}$ *and a singular measure* $\mathbf{P}_{si}$ *such that*

$$\mathbf{P} = c_1 \mathbf{P}_{at} + c_2 \mathbf{P}_{ct} + c_3 \mathbf{P}_{si},$$

where $c_1, c_2, c_3 \in [0, 1]$ *are such that* $c_1 + c_2 + c_3 = 1$.

Proof We already know that the atomic part of a measure is easy to distinguish on the basis of jumps (break points) of the distribution function. Let F be the distribution function of the variable X with $\mathcal{L}(X) = \mathbf{P}$. There are at most countable many such points $x \in \mathbb{R}$ for which $p_x = F(x_i^+) - F(x_i) > 0$. If $\sum_{x \in \mathbb{R}} p_x = c_1 > 0$, then

$$\mathbf{P}_{at} = \sum_{x \in \mathbb{R}} \frac{p_x}{c_1} \delta_x.$$

Let F_{at} be the distribution function of the measure $\mathbf{P}_{at}$. Then, $G(x) := F(x) - c_1 F_{at}(x)$ is a continuous non-decreasing function on $\mathbb{R}$, thus it is differentiable almost everywhere and

$$\int_{-\infty}^{z} G'(x) \, dx \leqslant F(z), \qquad \text{for all } z \in \mathbb{R}.$$

If $\int_{\mathbb{R}} G'(x) \, dx = c_2 > 0$, then the measure $\mathbf{P}_{ct}$ defined for Borel sets $B \in \mathcal{B}$ by the formula

$$\mathbf{P}_{ct}(B) = c_2^{-1} \int_B G'(x) \, dx$$

is a continuous type measure with density function $c_2^{-1} G'$. The singular part $\mathbf{P}_{si}$ of the distribution $\mathbf{P}$ is found by solving the following equation:

$$\mathbf{P} = c_1 \mathbf{P}_{at} + c_2 \mathbf{P}_{ct} + (1 - c_1 - c_2) \mathbf{P}_{si}.$$

□

3.6 Independent Random Variables

Definition 3.23 We say that random variables $X_1, \ldots, X_n$ are *independent* if for every choice of Borel sets $B_1, \ldots, B_n \in \mathcal{B}$, the random events $\{X_1 \in B_1\}, \ldots, \{X_n \in B_n\}$ are independent, i.e., the following equality holds:

$$\mathbf{P}\{X_1 \in B_1, \ldots, X_n \in B_n\} = \mathbf{P}\{X_1 \in B_1\} \cdot \ldots \cdot \mathbf{P}\{X_n \in B_n\}.$$

It is easy to see that if random variables $X_1, \ldots, X_n$ are independent and $s_1, \ldots, s_k \in \{1, \ldots, n\}$, $k < n$, $s_i \neq s_j$ for $i \neq j$, then the variables $X_{s_1}, \ldots, X_{s_k}$ are also independent. The next theorem states that functions defined on independent variables define independent variables.

Theorem 3.24 *Let us assume that random variables $X_1, \ldots, X_n$ are independent and the functions $\varphi_j : \mathbb{R} \to \mathbb{R}$, $j = 1, \ldots, n$, are Borel-measurable. Then, the random variables $Y_j = \varphi_j(X_j)$ are independent.*

Proof If φ_j, $j = 1, \ldots, n$, are Borel measurable functions, then for any Borel sets $B_1, \ldots, B_n$ the sets $\varphi_1^{-1}(B_1), \ldots, \varphi_n^{-1}(B_n)$ belong to the Borel σ-field as well, and

$$\begin{aligned}\mathbf{P}\{Y_1 \in B_1, \ldots, Y_n \in B_n\} &= \mathbf{P}\left\{X_1 \in \varphi_1^{-1}(B_1), \ldots, X_n \in \varphi_n^{-1}(B_n)\right\} \\ &= \mathbf{P}\left\{X_1 \in \varphi_1^{-1}(B_1)\right\} \ldots \mathbf{P}\left\{X_n \in \varphi_n^{-1}(B_n)\right\} \\ &= \mathbf{P}\{Y_1 \in B_1\} \ldots \mathbf{P}\{Y_n \in B_n\},\end{aligned}$$

which was to be shown. □

Theorem 3.25 *Assume that random variables $X_1, \ldots, X_n, X_{n+1}, \ldots, X_{n+m}$ are independent. If the functions $\varphi : \mathbb{R}^n \to \mathbb{R}$ and $\psi : \mathbb{R}^m \to \mathbb{R}$ are Borel-measurable, then the random variables $Y_1 = \varphi(X_1, \ldots, X_n)$ and $Y_2 = \psi(X_{n+1}, \ldots, X_{n+m})$ are also independent.*

Proof We need to check that for any choice of Borel sets A, B in $\mathbb{R}$ the following equality holds:

$$\mathbf{P}\{Y_1 \in A, Y_2 \in B\} = \mathbf{P}\{Y_1 \in A\} \mathbf{P}\{Y_2 \in B\}.$$

The set $C = \varphi^{-1}(A)$ is a Borel subset of $\mathbb{R}^n$ and $D = \psi^{-1}(B)$ is a Borel subset of $\mathbb{R}^m$, hence we need to show that for any Borel sets $C \subset \mathbb{R}^n$, $D \subset \mathbb{R}^m$, we have

$$\begin{aligned}\mathbf{P}\{(X_1, \ldots, X_n) &\in C, (X_{n+1}, \ldots, X_{n+m}) \in D\} \\ &= \mathbf{P}\{(X_1, \ldots, X_n) \in C\} \mathbf{P}\{(X_{n+1}, \ldots, X_{n+m}) \in D\}.\end{aligned}$$

It follows from the independence of the random variables $X_1, \ldots, X_{n+m}$ that this equality holds if the sets C and D are measurable rectangles, i.e., $C = C_1 \times \cdots \times C_n$ and $D = D_1 \times \cdots \times D_m$, where C_i and D_j are Borel subsets of the line.

Let $(\mathcal{K}, \mathcal{L})$ denote the set of all pairs (C, D), $C \subset \mathbb{R}^n$, $D \subset \mathbb{R}^m$, for which the desired equality holds. We already know that this set includes pairs of measurable rectangles. Since the σ-field in $\mathbb{R}^n$ is containing all measurable rectangles, hence it contains σ-field of Borel sets, and it is enough to prove that $\mathcal{K}$ and $\mathcal{L}$ are closed under countable set operations.

It is enough to prove that $\mathcal{K}$ is a σ-field. If $(C, D) \in (\mathcal{K}, \mathcal{L})$, then also $(C', D) \in (\mathcal{K}, \mathcal{L})$ because

$$\mathbf{P}\{(X_1, \ldots, X_n) \in \Omega \setminus C, (X_{n+1}, \ldots, X_{n+m}) \in D\}$$
$$= \mathbf{P}\{(X_1, \ldots, X_n) \in \Omega, (X_{n+1}, \ldots, X_{n+m}) \in D\}$$
$$- \mathbf{P}\{(X_1, \ldots, X_n) \in C, (X_{n+1}, \ldots, X_{n+m}) \in D\}$$
$$= \left(1 - \mathbf{P}\{(X_1, \ldots, X_n) \in C\}\right)\mathbf{P}\{(X_{n+1}, \ldots, X_{n+m}) \in D\}$$
$$= \mathbf{P}\{(X_1, \ldots, X_n) \in C'\}\mathbf{P}\{(X_{n+1}, \ldots, X_{n+m}) \in D\}.$$

In a similar way, we show that if $(C_k, D) \in (\mathcal{K}, \mathcal{L})$ for every $k \in \mathbb{N}$ and the events C_k are pairwise disjoint, then $(\bigcup C_k, D) \in (\mathcal{K}, \mathcal{L})$:

$$\mathbf{P}\left\{(X_1, \ldots, X_n) \in \bigcup C_n, (X_{n+1}, \ldots, X_{n+m}) \in D\right\}$$
$$= \mathbf{P}\left(\bigcup \{(X_1, \ldots, X_n) \in C_n, (X_{n+1}, \ldots, X_{n+m}) \in D\}\right)$$
$$= \sum_n \mathbf{P}\{(X_1, \ldots, X_n) \in C_n\}\mathbf{P}\{(X_{n+1}, \ldots, X_{n+m}) \in D\}$$
$$= \mathbf{P}\left\{(X_1, \ldots, X_n) \in \bigcup C_n\right\}\mathbf{P}\{(X_{n+1}, \ldots, X_{n+m}) \in D\}.$$

Now, let C and D be measurable rectangles. We already know that (C, D), $(C', D) \in (\mathcal{K}, \mathcal{L})$, hence $(C \cup C', D) = (\mathbb{R}^n, D) \in (\mathcal{K}, \mathcal{L})$. □

3.6.1 Exercises

184. Prove that if a random variable X is independent of any other random variable specified on the same probability space, then $X = c$ with probability one for some constant c.
185. Let $\Omega \subset \mathbb{R}$. Under what assumptions are the variables X and $\sin(X)$ independent?

3.6 Independent Random Variables

186. Random variables X, Y are independent with the same distribution given by the density $f(x)$. Find the densities of $Z = \max\{X, Y\}$ and $U = \min\{X, Y\}$.

187. Random variables X, Y, Z are independent with the same distribution given by the density $f(x)$. Find the distributions of the *ordered statistics* U_1, U_2, U_2, where

$$U_1(\omega) = \min\{X(\omega), Y(\omega), Z(\omega)\}, \quad U_3(\omega) = \max\{X(\omega), Y(\omega), Z(\omega)\};$$
$$U_2(\omega) = \{X(\omega), Y(\omega), Z(\omega)\} \setminus \{U_1(\omega), U_3(\omega)\}.$$

188. Let X, Y be independent random variables with distribution functions F and G, respectively. Find the distribution functions of the following random variables:

(a) $Z(\omega) = \max\{X(\omega), Y(\omega)\}$; (b) $W(\omega) = \min\{X(\omega), Y(\omega)\}$;
(c) $T(\omega) = \max\{2X(\omega), Y(\omega)\}$; (d) $U(\omega) = \min\{X^3(\omega), Y(\omega)\}$.

189. Random variables $X_1, X_2, \ldots$ are independent with the same distribution $\mathbf{P}\{X_i = 0\} = \mathbf{P}\{X_i = 1\} = \frac{1}{2}$. Find the distribution of the random variable $Y = \sum_{i=1}^{\infty} \frac{X_i}{2^i}$.

190. Variables X, Y are independent with the same uniform distribution on $[-1, 1]$. Find the distribution of the variable $Z = X + Y$.

191. Random variables $X_1, X_2, \ldots$ are independent with the same uniform distribution on the interval $[0, 1]$. For $\lambda > 0$, find the distribution of the variable

$$X = \inf\left\{n: \prod_{k=1}^{n} X_k < e^{-\lambda}\right\}.$$

192. The variable X_n is the sum of n independent random variables with uniform distributions on $[0, 1]$ and F_n is the cumulative distribution function of X_n. Prove that

$$F_{n+1}(x) = \int_{x-1}^{x} F_n(y)\,dy.$$

193. Let X and Y be independent with the same exponential distribution with parameter $\lambda > 0$. Find the distribution of the random variable $Z = X + Y$.

194. Let X and Y be independent with the same exponential distribution with parameter $\lambda > 0$. Find the distribution of the random variable $Z = X/Y$.

195. Let $N, X_1, X_2, \ldots$ be independent random variables with N having a geometric distribution $\mathbf{P}\{N = n\} = pq^{n-1}$, $n \geqslant 1$ and let each of the variables X_k have an exponential distribution $\Gamma(1, a)$. Show that the variable $Y = \sum_{k=1}^{N} X_k$ has exponential distribution.

196. The variable X has a standard normal distribution, and a random variable Y, independent of X, is uniformly distributed on the interval $[a, b]$. Find the density of the variable $Z = X + Y$.
197. The variables X, Y are independent with the same distribution given by the density $f(x) = (\pi \cosh(x))^{-1}$. Find the density of the variable $Z = X + Y$.

3.7 Multidimensional Random Variables and Distributions

Definition 3.26 A function $X : (\Omega, \mathcal{F}, \mathbf{P}) \mapsto \mathbb{R}^n$ is called an *n-dimensional random vector*, or *n-dimensional random variable* if, for every Borel set $B \subset \mathbb{R}^n$, the following condition holds:

$$\{\omega : (X_1(\omega), \ldots, X_n(\omega)) \in B\} \in \mathcal{F}.$$

The probability distribution of the random vector $X = (X_1, \ldots, X_n)$ is the probability measure $\mathbf{P}_X$ over the space $(\mathbb{R}^n, \mathcal{B}(\mathbb{R}^n))$ defined by the formula

$$\mathbf{P}_X(B) = \mathbf{P}\{\omega : (X_1(\omega), \ldots, X_n(\omega)) \in B\}, \quad B \in \mathcal{B}(\mathbb{R}^n).$$

It is easy to see that if $X = (X_1, \ldots, X_n)$ is a random vector, then each of the functions $X_i, i \leqslant n$, is a random variable.

As in the case of random variables, a simple random vector is a vector with a discrete distribution, while a continuous random vector is one with a density function. We will also consider convex linear combinations of the distributions of the previous two types. For the sake of simplicity, we will limit ourselves here to two-dimensional random vectors.

We say that a random vector (X, Y) has a *discrete distribution* if there exist sets $x_1, \ldots, x_n \in \mathbb{R}$ and $y_1, \ldots, y_m \in \mathbb{R}$ such that

$$\mathbf{P}\{X \in \{x_1, \ldots, x_n\}, Y \in \{y_1, \ldots, y_m\}\} = 1.$$

Let $p_{ij} = \mathbf{P}\{X = x_i, Y = y_j\}$. The joint distribution of the random variables X and Y, i.e., the distribution of the random vector (X, Y), can be described by the table:

$X \backslash Y$	y_1	y_2	$\ldots$	y_m	
x_1	$p_{1,1}$	$p_{1,2}$	$\ldots$	$p_{1,m}$	$p_{1\cdot}$
x_2	$p_{2,1}$	$p_{2,2}$	$\ldots$	$p_{2,m}$	$p_{2\cdot}$
$\ldots$	$\ldots$	$\ldots$	$\ldots$	$\ldots$	$\ldots$
x_n	$p_{n,1}$	$p_{n,2}$	$\ldots$	$p_{n,m}$	$p_{n\cdot}$
	$p_{\cdot 1}$	$p_{\cdot 2}$	$\ldots$	$p_{\cdot m}$	

3.7 Multidimensional Random Variables and Distributions

If the joint distribution of X and Y is known, then the distributions of X and Y are called *marginal distributions*. Let's find the marginal distribution of the random variable X:

$$p_{i\cdot} \stackrel{\text{def}}{=} \mathbf{P}\{X = x_i\} = \sum_{j=1}^{m} \mathbf{P}\{X = x_i, Y = y_j\} = \sum_{j=1}^{m} p_{i,j}.$$

The probability $p_{i\cdot}$ is, therefore, equal to the sum of all the elements of the i-th row of the matrix $(p_{i,j})$. Similarly,

$$p_{\cdot j} \stackrel{\text{def}}{=} \mathbf{P}\{Y = y_i\} = \sum_{i=1}^{n} \mathbf{P}\{X = x_i, Y = y_j\} = \sum_{i=1}^{n} p_{i,j},$$

so the probability $p_{\cdot j}$ is equal to the sum of all the elements of j-th column of the matrix $(p_{i,j})$.

If (X, Y) is a discrete random vector with distribution given by the matrix $(p_{i,j})$, then the conditional distribution of the variable X, assuming $Y = y_k$, is expressed by the formula:

$$\mathbf{P}\{X = x_i | Y = y_k\} = \frac{p_{i,k}}{p_{\cdot k}}.$$

Similarly, the conditional distribution of Y, assuming $X = x_i$, is given by the formula:

$$\mathbf{P}\{Y = y_k | X = x_i\} = \frac{p_{i,k}}{p_{i\cdot}}.$$

Let (X, Y) be a discrete random vector with probability matrix $(p_{i,j})$. Variables X and Y are independent if and only if

$$p_{i,j} = p_{i\cdot} p_{\cdot j} \quad \text{for all } i \in \{1, \ldots, n\}, \ j \in \{1, \ldots, m\}.$$

Definition 3.27 The *cumulative distribution function* of a random vector $X = (X_1, \ldots, X_n)$ is the function $F : \mathbb{R}^n \mapsto [0, 1]$ defined by the formula:

$$F(t_1, \ldots, t_n) = \mathbf{P}\{\omega : X_1(\omega) < t_1, \ldots, X_n(\omega) < t_n\}, \quad t_1, \ldots, t_n \in \mathbb{R}.$$

We say that a random vector (X, Y) is of *continuous type* if there exists an integrable non-negative function $f(x, y)$ on $\mathbb{R}^2$ such that for any Borel set $B \subset \mathbb{R}^2$

$$\mathbf{P}\{(X, Y) \in B\} = \int \ldots \int_{B} f(x, y) \, dx \, dy.$$

It can be shown that, as in the one-dimensional case, if at almost every point (x, y) the second partial derivative $\frac{\partial^2}{\partial x \partial y} F(x, y)$ of the two-dimensional distribution function $F(x, y)$ exists and the following condition holds

$$\int \cdots \int_{\mathbb{R}^2} \frac{\partial^2}{\partial x \partial y} F(x, y) \, dx \, dy = 1,$$

then the distribution of (X, Y) is absolutely continuous with density

$$f(x, y) = \frac{\partial^2}{\partial x \partial y} F(x, y).$$

Suppose $f(x, y)$ is the density function for the vector (X, Y). To find the marginal distribution of the variable X, let's first determine its distribution:

$$\mathbf{P}\{X < t\} = \mathbf{P}\{(X, Y) \in (-\infty, t) \times \mathbb{R}\} = \int_{-\infty}^{t} \int_{\mathbb{R}} f(x, y) \, dy \, dx.$$

It follows that the cumulative distribution function $F_X(t)$ of the variable X is equal to the integral over the set $(-\infty, t)$ of some integrable function. From this, we conclude that the variable X is of continuous type and its density is expressed by the formula

$$f_1(x) = \int_{\mathbb{R}} f(x, y) \, dy.$$

Similarly, we show that the marginal distribution of Y is also of continuous type with density

$$f_2(y) = \int_{\mathbb{R}} f(x, y) \, dx.$$

In this way, we have shown that if the random vector (X, Y) is of continuous type, then the distributions of X and Y are also of continuous type. Note that the converse implication does not apply. If X is a continuous type random variable with density f and we consider the random vector (X, X), then $\mathbf{P}\{(X, X) \in D\} = 1$, where $D = \{(x, y) : x = y\}$ is the diagonal of the coordinate system. Thus, (X, X) does not have a continuous type distribution, and both marginal distributions share this property.

The next theorem follows from the definition of a two-dimensional distribution function and the definition of independent random variables.

3.7 Multidimensional Random Variables and Distributions

Theorem 3.28 *Let $F_{X,Y}$ be a cumulative distribution function of a random vector (X, Y), and F_X and F_Y be the marginal distribution function of X and Y, respectively. The random variables X and Y are independent if and only if*

$$F_{X,Y}(t, s) = F_X(t) F_Y(s).$$

Note that if (X, Y) has a density $f(x, y)$, then the variables X and Y with densities f_1, f_2, respectively, are independent if and only if outside a set of Lebesgue measure zero, we have the equality:

$$f(x, y) = f_1(x) f_2(y).$$

Definition 3.29 If random variables X and Y are independent, X has distribution μ, and Y has distribution ν, then the distribution of the sum $X + Y$ is called the *convolution* of the distributions μ and ν, and we write:

$$\mathcal{L}(X + Y) = \mu * \nu.$$

If independent random variables X and Y are of continuous type with densities f and g, respectively, then the distribution function of the variable $X + Y$ has the form:

$$\mathbf{P}\{X + Y < t\} = \int \cdots \int_{x+y<t} f(x) g(y) \, dx \, dy$$

$$= \int_{-\infty}^{t} \int_{\mathbb{R}} f(x - y) g(y) \, dy \, dx.$$

We conclude from this that the convolution of these two distributions is a continuous distribution and has a density $h(x)$ given by the formula:

$$h(x) = \int_{\mathbb{R}} f(x - y) g(y) \, dy.$$

By analogy with the convolution of distributions, we then say that h is the *convolution of the densities* f and g, which we denote by:

$$h(x) = f * g(x) \stackrel{\text{def}}{=} \int_{\mathbb{R}} f(x - y) g(y) \, dy.$$

Suppose (X, Y) has a continuous type distribution with a uniformly continuous density function $f(\cdot, \cdot)$. Let us informally define a random variable $(X|Y = y)$ which takes the value of $X(\omega)$ provided that $Y(\omega) = y$.

The density of the variable $(X|Y = y)$ under the very strong assumption of uniform integrability of the density f is expressed by the formula:

$$f_{X|Y}(x|y) = \begin{cases} \frac{f(x,y)}{\int_{\mathbb{R}} f(x,y)\,dx} & \text{if } \int_{\mathbb{R}} f(x,y)\,dx \neq 0; \\ 0 & \text{otherwise.} \end{cases}$$

To see this, it suffices to note that

$$\mathbf{P}\{X \in B | y - \varepsilon < Y < y + \varepsilon\} = \frac{\int_B \int_{y-\varepsilon}^{y+\varepsilon} f(x,y)\,dy\,dx}{\int_{\mathbb{R}} \int_{y-\varepsilon}^{y+\varepsilon} f(x,y)\,dy\,dx}$$

and then take the limit as $\varepsilon \to 0$.

3.7.1 Exercises

198. A two-dimensional random variable (X, Y) has a discrete distribution given by the following conditions:

$$\mathbf{P}\{X = 1, Y = 1\} = \mathbf{P}\{X = 1, Y = 2\} = \mathbf{P}\{X = 2, Y = 2\} = \frac{1}{3}.$$

Find the matrix of the two-dimensional distribution, two-dimensional cumulative distribution function and marginal distribution functions. Are the variables X and Y independent?

199. Consider independent random variables with the following distributions:

$$\mathbf{P}\{X = 1\} = 0.5; \quad \mathbf{P}\{X = 2\} = \mathbf{P}\{X = 3\} = 0.25;$$
$$\mathbf{P}\{Y = 0\} = 0.5; \quad \mathbf{P}\{Y = 1\} = \mathbf{P}\{Y = 2\} = 0.25.$$

Find the joint distribution of the random vector (X, Y).

200. The two-dimensional density function is given by:

$$f(x, y) = \begin{cases} e^{-x-y} & \text{if } x > 0, y > 0; \\ 0 & \text{otherwise.} \end{cases}$$

Calculate: $\mathbf{P}\{1 < X < 2, 1 < Y < 2\}$, $\mathbf{P}\{X + Y > 2\}$, $\mathbf{P}\{X > 3 | Y < 1\}$. Find the two-dimensional cumulative distribution function and the marginal distribution functions. Are the variables X and Y independent?

201. Is it possible to choose the constant C such that the function:

$$f(x, y) = \begin{cases} Cy^2 \cos x & \text{if } \frac{\pi}{2} < x < \pi, 0 < y < 2; \\ 0 & \text{otherwise} \end{cases}$$

is the density of a two-dimensional random variable? If so, find the marginal density functions.

202. Let α, β be random variables such that both real roots of the quadratic equation $x^2 + \alpha x + \beta = 0$ can independently take any value from the interval $[-1, 1]$. Find the distributions of the variables α and β.

203. Let X and Y be independent random variables with the same exponential distribution with parameter $\alpha > 0$. Find the distribution functions and the densities for the following random variables:

(a) $S(\omega) = \max\{X(\omega), Y^3(\omega)\}$; (b) $T(\omega) = X(\omega) + Y(\omega)$;
(c) $U(\omega) = 3 + 2X(\omega)$; (d) $V(\omega) = |X(\omega) - Y(\omega)|$;
(e) $W(\omega) = X^2(\omega)$; (f) $Z(\omega) = X/Y$.

204. Let X be a random variable with a) a Cauchy distribution; b) an exponential distribution. Let Θ be independent of the random variable X such that $\mathbf{P}\{\Theta = 1\} = 1 - \mathbf{P}\{\Theta = -1\} = p$ for some $p \in (0, 1)$. Find the distribution function and the density for the random variable $Y = X \cdot \Theta$.

205. A random vector (X, Y) has density $f(x, y)$. Find the distribution functions and densities for the following variables:

(a) $Z = X+Y$; (b) $U = X-Y$; (c) $W = X \cdot Y$; d) $U = X/Y$ if $\mathbf{P}\{Y \neq 0\} = 1$.

Find the densities of these variables if X and Y are independent.

206. A two-dimensional random vector (X, Y) has density function $f(x, y)$. Find the distribution function and the density of the variable $Z = X/(X+Y)$. Find the density of the variable Z under the assumption that X and Y are independent.

207. Random variables X and Y are independent with the same exponential distribution with parameter $\lambda = 2$. Calculate $\mathbf{P}\{X < Y < t\}$, where $t > 0$.

208. Random variables X and Y are independent and have the same distribution given by the density f. Calculate $\mathbf{P}\{X > Y\}$.

209. Independent random variables X and Y have the same discrete geometric distribution with parameter $p \in (0, 1)$. Find the distribution of the variable $Z = X + Y$.

210. Let X, Y be independent Poisson random variables with parameters λ_1, λ_2, respectively. Find the distribution of the variable $Z = X + Y$.

211. The probability that k job-seekers will report to an employment agency within an hour is equal to $\frac{\lambda^k}{k!}e^{-\lambda}$, where $\lambda > 0$ is a parameter. For each of these people, the probability of finding a job is p. Find the probability that exactly n of the people who applied between 10.00 a.m. and 11.00 a.m. will find a job.

212. Random variables X and Y are independent, X has an exponential distribution with parameter λ, and Y is uniformly distributed on the interval $[0, h]$. Find the distribution density of the following variables: a) $Z = X+Y$; b) $W = X - Y$.

213. A random variable X has a two-point distribution. Prove that X cannot be represented as the sum of any two independent random variables.
214. A random variable X takes the values $-1, 0, 1$ with probabilities p_1, p_2, p_3, respectively, where $p_1 p_2 p_3 > 0$ and $p_1 + p_2 + p_3 = 1$. What conditions must be met by the numbers p_1, p_2, p_3 for X to be represented as the sum of two independent random variables with the same distributions?
215. Prove that the convolution of two discrete distributions is also discrete.
216. Prove that $X + Y$ has a continuous type distribution if at least one of the independent variables X or Y has a continuous type distribution.
217. X and Y are independent random variables with distributions $\Gamma(p, a)$ and $\Gamma(q, a)$, respectively. Find the distribution of the variable $Z = X + Y$.
218. X and Y are independent random variables with uniform distributions on the intervals $[a, b]$ and $[c, d]$, respectively, where $0 < a < b < c < d$. Find the distribution of the variable $Z = X + Y$.
219. A random vector (X, Y, Z) has density function

$$f(x, y, z) = \begin{cases} 6(1 + x + y + z)^{-4} & \text{if } x, y, z > 0; \\ 0 & \text{otherwise.} \end{cases}$$

Find the distribution of the random variable $W = X + Y + Z$.
220. Prove that if random variables X, Y are independent with the same exponential distribution with parameter 1, then the variables $Z = X + Y$ and $W = X/Y$ are also independent.
221. Prove that if random variables X, Y are independent with the same normal distribution $\mathcal{N}(0, \sigma)$, then the variables $Z = X^2 + Y^2$ and $W = X/Y$ are also independent.
222. Prove that if random variables X, Y are independent, X has the distribution $\Gamma(p, a)$, and Y has the distribution $\Gamma(q, a)$, then the variables $Z = X + Y$ and $W = X/Y$ are also independent.

Chapter 4
Expected Value for Random Variables

4.1 Expected Value for Simple Random Variables

Let us consider one of the simplest games of chance: a player rolls a single cubical die. If the result is a 6 he wins a euros but for any of the other possible outcomes he loses b euros. On average, what kind of winnings can he expect?

The classic lottery game—Totolotek—involves choosing six out of 49 numbers that you hope will be randomly selected in the draw. The probability of losing is very high, but you can also win, and the more numbers you match, the higher your winnings. What kind of winnings can we expect on average?

Such problems have been studied since the very beginning of probability theory. The first records date back to the beginning of the eighteenth century. It was then that the concept of *mathematical hope* was introduced. Mathematical hope has also been called *average value, expected value, mean value* or *esperance*.

We begin here by defining the expected value for simple random variables. Let $(\Omega, \mathcal{F}, \mathbf{P})$ be a probability space and let $X : \Omega \to \mathbb{R}$ be a simple variable (simple function). That is, there are the numbers $x_1, \ldots, x_n \in \mathbb{R}$ and sets $A_1, \ldots, A_n \in \mathcal{F}$ such that

$$\forall i \neq j \; A_i \cap A_j = \emptyset, \quad \bigcup_{k=1}^{n} A_k = \Omega \quad \text{and} \quad X(\omega) = \sum_{k=1}^{n} x_k \mathbf{1}_{A_k}(\omega).$$

Then, the *expected value* $\mathbf{E}X$ of the variable X is defined by the formula:

$$\mathbf{E}X \stackrel{\text{def}}{=} \sum_{k=1}^{n} x_k \mathbf{P}(A_k).$$

Instead of the expected value symbol $\mathbf{E}X$, sometimes we will use the notation $\int_\Omega X(\omega)\mathbf{P}(d\omega)$ or $\int_\Omega X\, d\mathbf{P}$ to emphasize that the expected value is the integral of $X(\omega)$ with respect to the probability measure $\mathbf{P}$.

Lemma 4.1 *The expected value of a simple random variable is uniquely defined.*

Proof Suppose that a simple random variable X can be written in two ways: $X(\omega) = \sum_{k=1}^{n} x_k \mathbf{1}_{A_k}(\omega) = \sum_{j=1}^{m} y_j \mathbf{1}_{B_j}(\omega)$, where $A_i \cap A_j = \emptyset$ and $B_i \cap B_j = \emptyset$ if and only if $i \neq j$ and at the same time $\bigcup_{k=1}^{n} A_k = \bigcup_{j=1}^{m} B_j = \Omega$. Note that

$$A_k = A_k \cap \Omega = A_k \cap \bigcup_{j=1}^{m} B_j = \bigcup_{j=1}^{m} (A_k \cap B_j).$$

So, if $\omega \in A_k \cap B_j \neq \emptyset$, then $X(\omega) = x_k = y_j$. This implies that

$$\sum_{k=1}^{n} x_k \mathbf{P}(A_k) = \sum_{k=1}^{n} x_k \mathbf{P}\left(\bigcup_{j=1}^{m}(A_k \cap B_j)\right) = \sum_{k=1}^{n} x_k \sum_{j=1}^{m} \mathbf{P}(A_k \cap B_j)$$

$$= \sum_{j=1}^{m} \sum_{k=1}^{n} x_k \mathbf{P}(A_k \cap B_j) = \sum_{j=1}^{m} \sum_{k=1}^{n} y_j \mathbf{P}(A_k \cap B_j)$$

$$= \sum_{j=1}^{m} y_j \mathbf{P}\left(\bigcup_{k=1}^{n}(A_k \cap B_j)\right) = \sum_{j=1}^{m} y_j \mathbf{P}(B_j).$$

$\square$

Remark 4.2 The previous lemma shows that for any simple random variable X taking values $x_1, \ldots, x_n$ with probabilities $p_1, \ldots, p_n$, it can be assumed that $A_k = \{\omega : X(\omega) = x_k\}$. Hence:

$$\mathbf{E}X = \sum_{k=1}^{n} x_k \mathbf{P}(A_k) = \sum_{k=1}^{n} x_k\, p_k.$$

In a similar way, we get for $r \in \mathbb{N}$:

$$\mathbf{E}X^r = \sum_{k=1}^{n} x_k^r\, \mathbf{P}\{X^r = x_k^r\} = \sum_{k=1}^{n} x_k^r\, \mathbf{P}\{X = x_k\} = \sum_{k=1}^{n} x_k^r\, p_k.$$

Hence, if, when rolling a single die, receiving a six-pip result wins a, and any other result loses b, the mathematical hope in this game is $(a - 5b)/6$. If, on the other hand, X stands for the number of pips obtained, we have that $\mathbf{E}X = \frac{1}{6}(1 + 2 + 3 + 4 + 5 + 6) = 3.5$.

4.1 Expected Value for Simple Random Variables

Examples 4.3

1. Let the random variable X have a hypergeometric distribution with parameters $N, M, n, n \leqslant N, n \leqslant M$, i.e.,

$$p_k = \mathbf{P}\{X = k\} = \frac{\binom{N}{k}\binom{M}{n-k}}{\binom{M+N}{n}}, \quad k = 0, 1, \ldots, n.$$

Since for every discrete probability distribution $\sum_{k=0}^{n} p_k = 1$, we get the following combinatorial identity:

$$\sum_{k=0}^{n} \binom{N}{k}\binom{M}{n-k} = \binom{N+M}{n}.$$

Now, we can calculate the expected value for X:

$$\mathbf{E}X = \sum_{k=0}^{n} k \frac{\binom{N}{k}\binom{M}{n-k}}{\binom{N+M}{n}}$$

$$= \frac{N}{\binom{N+M}{n}} \sum_{k=1}^{n} \binom{N-1}{k-1}\binom{M}{n-k}$$

$$= \frac{N}{\binom{N+M}{n}} \binom{N+M-1}{n-1}$$

$$= \frac{nN}{N+M}.$$

2. Let $\Omega = [0, 1]$ and let $\mathbf{P}$ be the Lebesgue measure on Ω. The random variable $X(\omega)$ takes the value 1 if ω is a rational number and it takes 0 otherwise. It is therefore a simple random variable that takes only two values. To find its expected value, note that the Lebesgue measure of any point is equal to zero and we only have countably many rational numbers. Hence,

$$\mathbf{P}\{\omega : X(\omega) = 1\} = \mathbf{P}(Q) = \mathbf{P}\left(\bigcup_{r \in Q}\{r\}\right) = \sum_{r \in Q} \mathbf{P}(\{r\}) = 0.$$

We get

$$\mathbf{E}X = \int_0^1 \left(1 \cdot \mathbf{1}_Q(\omega) + 0 \cdot \mathbf{1}_{Q'}(\omega)\right) \mathbf{P}(d\omega)$$

$$= 1 \cdot \mathbf{P}(Q) + 0 \cdot \mathbf{P}(Q') = 0.$$

It can be seen that the considered integral with respect to the probability measure (in this case the Lebesgue measure), called the expected value, is significantly different from the Riemann integral, defined by the common limit of the lower sums $\underline{s_n}$ and the upper sums $\overline{s_n}$, where

$$\underline{s_n} = \sum_{k=1}^{n} m_k \mathbf{P}(\Delta_k), \qquad \overline{s_n} = \sum_{k=1}^{n} M_k \mathbf{P}(\Delta_k),$$

and $\Delta_1, \ldots, \Delta_n$ is a partition of the interval $[0, 1]$ into disjoint intervals with lengths going down to zero, m_k being the smallest, and M_k being the largest value of the function $X(\omega)$ on the interval Δ_k. In our case, $m_k = 0$ and $M_k = 1$ for each k, hence $\underline{s_n} = 0$ and $\overline{s_n} = 1$ for each n, which means that the Riemann integral of the variable $X(\omega)$ as the joint limit of $\underline{s_n}$ and $\overline{s_n}$ does not exist.

Lemma 4.4 *If X and Y are simple random variables, then*

(1) $X = \mathbf{1}_A(\omega) \Longrightarrow \mathbf{E}X = \mathbf{P}(A)$;
(2) $X \geqslant 0 \Longrightarrow \mathbf{E}X \geqslant 0$;
(3) $\forall a, b \in \mathbb{R} \ \ \mathbf{E}(aX + bY) = a\mathbf{E}X + b\mathbf{E}Y$;
(4) $X \geqslant Y \Longrightarrow \mathbf{E}X \geqslant \mathbf{E}Y$;
(5) $|\mathbf{E}X| \leqslant \mathbf{E}|X|$;
(6) *if X and Y are independent, then $\mathbf{E}(X \cdot Y) = \mathbf{E}X \cdot \mathbf{E}Y$.*

Proof The first two properties are obvious and property 4 is a simple consequence of properties 2 and 3, so we only prove the other properties.

(3) Let $X = \sum_{k=1}^{n} x_k \mathbf{1}_{A_k}$ and $Y = \sum_{j=1}^{m} y_j \mathbf{1}_{B_j}$. Then

$$aX + bY = a \sum_{j,k} x_k \mathbf{1}_{A_k \cap B_j} + b \sum_{j,k} y_j \mathbf{1}_{A_k \cap B_j} = \sum_{j,k} (ax_k + by_j) \mathbf{1}_{A_k \cap B_j}.$$

Hence, we have

$$\mathbf{E}(aX + bY) = \sum_{j,k} (ax_k + by_j) \mathbf{P}(A_k \cap B_j)$$

$$= a \sum_{j,k} x_k \mathbf{P}(A_k \cap B_j) + b \sum_{j,k} y_j \mathbf{P}(A_k \cap B_j)$$

$$= a \sum_k x_k \sum_j \mathbf{P}(A_k \cap B_j) + b \sum_j y_j \sum_k \mathbf{P}(A_k \cap B_j)$$

$$= a \sum_k x_k \mathbf{P}(A_k) + b \sum_j y_j \mathbf{P}(B_j) = a\mathbf{E}X + b\mathbf{E}Y.$$

4.1 Expected Value for Simple Random Variables

(5) It suffices to note that

$$|\mathbf{E}X| = \left|\sum_{k=1}^{n} x_k \mathbf{P}(A_k)\right| \leq \sum_{k=1}^{n} |x_k| \mathbf{P}(A_k) = \mathbf{E}|X|.$$

(6) From Lemma 4.1, it follows that from all possible representations of X and Y, without losing generality, we can consider those for which $A_k = \{\omega : X(\omega) = x_k\}$, $B_j = \{\omega : Y(\omega) = y_j\}$. Then,

$$\mathbf{P}(A_k \cap B_j) = \mathbf{P}\{\omega : X(\omega) = x_k, Y(\omega) = y_j\}$$
$$= \mathbf{P}\{\omega : X(\omega) = x_k\} \mathbf{P}\{\omega : Y(\omega) = y_j\} = \mathbf{P}(A_k) \mathbf{P}(B_j).$$

Hence, we get:

$$\mathbf{E}(XY) = \mathbf{E}\left(\sum_{j,k} x_k y_j \mathbf{1}_{A_k \cap B_j}\right) = \sum_{k=1}^{n}\sum_{j=1}^{m} x_k y_j \mathbf{P}(A_k \cap B_j)$$
$$= \sum_{k=1}^{n}\sum_{j=1}^{m} x_k y_j \mathbf{P}(A_k) \mathbf{P}(B_j) = \sum_{k=1}^{n} x_k \mathbf{P}(A_k) \cdot \sum_{j=1}^{m} y_j \mathbf{P}(B_j) = \mathbf{E}X \cdot \mathbf{E}Y.$$

□

4.1.1 Exercises

223. Calculate $\mathbf{E}X$ and $\mathbf{E}X^2$ for a random variable X with the following distributions: (a) one-point; (b) two-point; and (c) binomial with parameters n and p.
224. The random variable X is the number of pips thrown in a roll of a die. Calculate $\mathbf{E}X$ and $\mathbf{E}X^2$.
225. Will and Paul decided to play the following game: if the single roll of a die results in 1 or 2, Paul receives a euros from Will; for any other result, Paul pays Will b euros. What should be the relationship between a and b for a fair game?
226. Assume that a non-negative random variable X takes values in $\{a_1, \ldots, a_k\}$ and $\mathbf{P}\{X = a_j\} > 0$ for each $j \leq n$. Prove that

$$\lim_{n \to \infty} \frac{\mathbf{E}X^{n+1}}{\mathbf{E}X^n} = \lim_{n \to \infty} \sqrt[n]{\mathbf{E}X^n} = \max\{a_1, \ldots, a_k\}.$$

227. What is the expected value of the number of aces in your hand (13 cards) at the beginning of a bridge game?

228. The number of winning tickets in a lottery is N and the number of losing tickets is M. I have bought k of them. Let X be the number of winning tickets among those I have bought. Calculate $\mathbf{E}X$.

4.2 General Definition of Expected Value

We will need the following two lemmas to define the expected value for non-negative random variables.

Lemma 4.5 *If X is a non-negative random variable, then there exists a nondecreasing sequence of simple non-negative random variables (X_n) such that*

$$\forall \omega \in \Omega \quad X_n(\omega) \nearrow X(\omega).$$

Proof All we need is to define the sequence $\{X_n\}$. We can do this, for example, in the following way:

$$X_n(\omega) = \begin{cases} \dfrac{k}{2^n} & \text{if } \dfrac{k}{2^n} \leq X(\omega) < \dfrac{k+1}{2^n}, \ k = 0, 1, \ldots, n2^n - 1; \\ n & \text{if } X(\omega) \geq n. \end{cases}$$

With the notation

$$A_{k,n} = \left\{ \omega : \dfrac{k}{2^n} \leq X(\omega) < \dfrac{k+1}{2^n} \right\},$$

$$A_n = \{\omega : X(\omega) \geq n\},$$

the random variable X_n can be written as:

$$X_n = \sum_{k=0}^{n2^n-1} \dfrac{k}{2^n} \mathbf{1}_{A_{k,n}} + n \mathbf{1}_{A_n}.$$

Of course, for each $\omega \in \Omega$ and every $n \in \mathbb{N}$, we have the following inequalities: $X_n(\omega) \leq X_{n+1}(\omega) \leq X(\omega)$. If for a fixed $\omega \in \Omega$ there exists an n_0 such that $X(\omega) < n_0$, then for every $n \geq n_0$, we have

$$0 \leq X(\omega) - X_n(\omega) < \dfrac{1}{2^n} \xrightarrow{n \to \infty} 0.$$

If such an n_0 did not exist, then we would have $X(\omega) = +\infty$, which is impossible because, by definition, every random variable takes values in the open line $\mathbb{R}$. □

4.2 General Definition of Expected Value

Lemma 4.6 *If Y and X_n, $n = 1, 2, \ldots$, are non-negative simple random variables and for each $\omega \in \Omega$ we have $X_n(\omega) \nearrow X(\omega) \geqslant Y(\omega)$, then*

$$\lim_{n \to \infty} \mathbf{E} X_n \geqslant \mathbf{E} Y.$$

Proof Let $\varepsilon > 0$ and $A_n = \{\omega : X_n(\omega) \geqslant Y(\omega) - \varepsilon\}$. Since $X_{n+1}(\omega) \geqslant X_n(\omega)$, the sequence of sets A_n is increasing ($A_n \subset A_{n+1}$). As at the same time $X_n(\omega) \nearrow X(\omega) \geqslant Y(\omega)$, for every $\omega \in \Omega$ there exists an $n_0 \in \mathbb{N}$ such that $\omega \in A_n$ for every $n \geqslant n_0$. Hence, we get:

$$\Omega = \bigcup_{n=1}^{\infty} A_n \quad \text{and} \quad \mathbf{P}(A_n) \xrightarrow{n \to \infty} 1.$$

Note that $X_n \mathbf{1}_{A_n}$ is a simple function since X_n is a simple function and

$$X_n = X_n \mathbf{1}_{A_n} + X_n \mathbf{1}_{A'_n} \geqslant X_n \mathbf{1}_{A_n} \geqslant (Y - \varepsilon) \mathbf{1}_{A_n}.$$

A simple function takes only finite many values, thus, if we denote by a the maximum value of the function $Y \mathbf{1}_{A'_n}$, we get:

$$\begin{aligned} \mathbf{E} X_n &\geqslant \mathbf{E}\left[(Y - \varepsilon) \mathbf{1}_{A_n}\right] = \mathbf{E}\left[Y \mathbf{1}_{A_n}\right] - \varepsilon \mathbf{P}(A_n) \\ &= \mathbf{E} Y - \mathbf{E}\left[Y \mathbf{1}_{A'_n}\right] - \varepsilon \mathbf{P}(A_n) \\ &\geqslant \mathbf{E} Y - \max\left\{Y(\omega) : \omega \in A'_n\right\} \mathbf{P}(A'_n) - \varepsilon \mathbf{P}(A_n) \\ &= \mathbf{E} Y - a \mathbf{P}(A'_n) - \varepsilon \mathbf{P}(A_n). \end{aligned}$$

Since $\mathbf{P}(A'_n) \to 0$ and $\mathbf{P}(A_n) \to 1$ for $n \to \infty$, we have

$$\lim_{n \to \infty} \mathbf{E} X_n \geqslant \mathbf{E} Y - \varepsilon.$$

The result follows by passing to the limit as $\varepsilon \to 0$. □

Definition 4.7 *If $X \geqslant 0$ and (X_n), $n = 1, 2, \ldots$, is a sequence of non-negative simple functions such that for every $\omega \in \Omega$ we have $X_n(\omega) \nearrow X(\omega)$, then the expected value for the variable X is defined by*

$$\mathbf{E} X \equiv \int_\Omega X(\omega) \mathbf{P}(d\omega) \stackrel{\text{def}}{=} \lim_{n \to \infty} \int_\Omega X_n(\omega) \mathbf{P}(d\omega) \equiv \lim_{n \to \infty} \mathbf{E} X_n.$$

The sequence (X_n) is called the *supporting* or *approximating sequence* of simple functions for the non-negative X.

Remark 4.8 It is easy to see that for a random variable X which takes values in $\{0, 1, 2, \ldots\}$ an approximating sequence of simple variables can be:

$$X_n = \sum_{k=0}^{n-1} k \mathbf{1}_{\{X=k\}} + n \mathbf{1}_{\{X \geqslant n\}}.$$

Consequently,

$$\mathbf{E}X = \lim_{n \to \infty} \mathbf{E}X_n = \sum_{k=0}^{\infty} k \mathbf{P}\{X = k\}.$$

Examples 4.9

1. If X has a discrete geometric distribution with $p \in (0, 1)$ (which is the waiting time for the first success in a sequence of Bernoulli trials), then

$$\mathbf{E}X = \sum_{k=1}^{\infty} k p q^{k-1} = p \left(\sum_{k=1}^{\infty} q^k \right)'_{dq} = p \left(\frac{q}{1-q} \right)'_{dq} = \frac{1}{p}.$$

2. If a random variable X has a Pascal distribution (which is the waiting time for the r-th success) with parameters $p \in (0, 1)$, $r \in \mathbb{N}$, then its expected value can be obtained as the sum:

$$\mathbf{E}X = \sum_{k=r}^{\infty} k \binom{k-1}{r-1} p^r q^{k-r},$$

which may seem rather difficult. It is easier to see that the waiting time for the r-th success is the sum of the waiting times for subsequent successes, hence $X = \sum_{j=1}^{r} X_j$, where $X_1, \ldots, X_r$ are independent random variables of the same geometric distribution with parameter p. Hence, we get:

$$\mathbf{E}X = \sum_{j=1}^{r} \mathbf{E}X_j = \frac{r}{p}.$$

3. Assume that X has a Poisson distribution with parameter $\lambda > 0$. Then:

$$\mathbf{E}X = \sum_{k=0}^{\infty} k e^{-\lambda} \frac{\lambda^k}{k!} = \lambda e^{-\lambda} \sum_{k=1}^{\infty} \frac{\lambda^{k-1}}{(k-1)!} = \lambda.$$

4.2 General Definition of Expected Value

4. Consider a random variable $X(\omega) = \omega$ on the probability space $\Omega = [0, 1]$ with the geometric probability, i.e., **P** is the Lebesgue measure. An approximating sequence of simple variables can be defined similarly to the proof of Lemma 4.5:

$$X_n(\omega) = \sum_{k=0}^{n-1} \frac{k}{n} \cdot \mathbf{1}_{[\frac{k}{n}, \frac{k+1}{n})}.$$

Thus, we have

$$\mathbf{E}X = \lim_{n \to \infty} \mathbf{E}X_n = \lim_{n \to \infty} \sum_{k=0}^{n-1} \frac{k}{n} \cdot \mathbf{P}\left(\left[\frac{k}{n}, \frac{k+1}{n}\right)\right) = \lim_{n \to \infty} \sum_{k=0}^{n-1} \frac{k}{n} \frac{1}{n} = \frac{1}{2}.$$

Lemma 4.10 *The expected value for non-negative random variables is well defined, i.e., it does not depend on the choice of the approximating sequence of simple variables.*

Proof Consider two strings (X_n), (Y_n) of simple, non-negative variables such that $X_n(\omega) \nearrow X(\omega)$, $Y_n(\omega) \nearrow X(\omega)$ for every $\omega \in \Omega$. Since $X_n \nearrow X \geqslant Y_k$, from Lemma 4.6, we get

$$\forall k \in \mathbb{N} \quad \lim_{n \to \infty} \mathbf{E}X_n \geqslant \mathbf{E}Y_k$$

and, consequently,

$$\lim_{n \to \infty} \mathbf{E}X_n \geqslant \lim_{k \to \infty} \mathbf{E}Y_k.$$

By changing roles of the variables X_n and Y_k from the same lemma, we get the converse inequality. □

Now, we can define the expected value for any random variable X on $(\Omega, \mathcal{F}, \mathbf{P})$. We introduce the following notation:

$$X^+(\omega) = \begin{cases} X(\omega) & \text{if } X(\omega) > 0; \\ 0 & \text{if } X(\omega) \leqslant 0; \end{cases} \quad X^-(\omega) = \begin{cases} -X(\omega) & \text{if } X(\omega) < 0; \\ 0 & \text{if } X(\omega) \geqslant 0. \end{cases}$$

Definition 4.11 The *expected value of a random variable* X is defined by the formula:

$$\mathbf{E}X \stackrel{\text{def}}{=} \mathbf{E}X^+ - \mathbf{E}X^-$$

if at least one of the two values on the right is finite. Otherwise, we say that the expected value for such variable X does not exist.

Theorem 4.12 *If the expected value* $\mathbf{E}X$ *exists and* $c \in \mathbb{R}$, *then* $\mathbf{E}(cX)$ *also exists and* $\mathbf{E}(cX) = c\,\mathbf{E}X$.

Proof If $X \geqslant 0$, $c \geqslant 0$ and (X_n) is a sequence of non-negative simple random variables such that $X_n \nearrow X$, then

$$\mathbf{E}(cX) = \lim_{n \to \infty} \mathbf{E}(cX_n) = c \lim_{n \to \infty} \mathbf{E}X_n = c\mathbf{E}X.$$

Now, it is easy to prove that the result holds for any integrable random variable X. We need to apply the general definition of the expected value bearing in mind that for $c > 0$, we have $(cX)^+ = c\,X^+$, $(cX)^- = c\,X^-$, and if $c < 0$ then $(cX)^+ = -c\,X^-$, $(cX)^- = -c\,X^+$. □

Theorem 4.13 *If* $\mathbf{E}|X| < \infty$ *and* $\mathbf{E}|Y| < \infty$, *then the expected value* $\mathbf{E}(X+Y)$ *exists and* $\mathbf{E}(X+Y) = \mathbf{E}X + \mathbf{E}Y$.

Proof

(1) Note first that if $X \geqslant 0$ and $Y \geqslant 0$, and the sequences of simple functions (X_n) and (Y_n) support X and Y, respectively, then $(X_n + Y_n)$ are also simple variables, and $(X_n + Y_n) \nearrow (X + Y)$. Hence, we obtain

$$\mathbf{E}X + \mathbf{E}Y = \lim_{n \to \infty} \mathbf{E}X_n + \lim_{n \to \infty} \mathbf{E}Y_n = \lim_{n \to \infty} \mathbf{E}(X_n + Y_n) = \mathbf{E}(X+Y).$$

(2) Assume now that X and Y are non-negative random variables, $\mathbf{E}|X| < \infty$ and $\mathbf{E}|Y| < \infty$. Then,

$$(X+Y)^+ = \frac{1}{2}(|X+Y| + X + Y) \leqslant \frac{1}{2}(|X| + |Y| + X + Y) = X^+ + Y^+.$$

It follows that

$$\mathbf{E}(X+Y)^+ \leqslant \mathbf{E}(X^+ + Y^+) = \mathbf{E}X^+ + \mathbf{E}Y^+ < \infty,$$

which guarantees the existence of the expected value for the variable $(X+Y)$. In the same way it can be shown that $\mathbf{E}(X+Y)^- < \infty$, hence $\mathbf{E}|X+Y| < \infty$. Now let us note that

$$(X+Y)^+ + X^- + Y^- = \frac{1}{2}(|X+Y| + X + Y) + X^- + Y^-$$

$$= \frac{1}{2}(|X+Y| + |X| + |Y|)$$

4.2 General Definition of Expected Value

and

$$(X+Y)^- + X^+ + Y^+ = -\left[X + Y - \frac{1}{2}(|X+Y| + X + Y)\right] + X^+ + Y^+$$

$$= \frac{1}{2}(-X - Y + |X+Y| + 2X^+ + 2Y^+) = \frac{1}{2}(|X+Y| + |X| + |Y|).$$

Hence, we obtain

$$\mathbf{E}(X+Y)^+ + \mathbf{E}X^- + \mathbf{E}Y^- = \mathbf{E}(X+Y)^- + \mathbf{E}X^+ + \mathbf{E}Y^+,$$

and, after rearranging,

$$\mathbf{E}(X+Y)^+ - \mathbf{E}(X+Y)^-$$
$$= \left[\mathbf{E}X^+ - \mathbf{E}X^-\right] + \left[\mathbf{E}Y^+ - \mathbf{E}Y^-\right] = \mathbf{E}X + \mathbf{E}Y.$$

□

Theorem 4.14 *If there exist expected values $\mathbf{E}X$ and $\mathbf{E}Y$ and $X(\omega) \geqslant Y(\omega)$ for each $\omega \in \Omega$, then $\mathbf{E}X \geqslant \mathbf{E}Y$.*

Proof Since $X - Y \geqslant 0$, we have $\mathbf{E}(X - Y) \geqslant 0$ as the limit of expected values of non-negative approximating simple variables. Thus, if both expected values are finite, we get $\mathbf{E}X - \mathbf{E}Y \geqslant 0$. In the other cases, we obtain the result just as easily by noting that $X^+ \geqslant Y^+$ and $X^- \leqslant Y^-$. □

Theorem 4.15

(a) *The expected value $\mathbf{E}X$ is finite if and only if $\mathbf{E}|X| < \infty$.*
(b) *If $\mathbf{E}|X| < \infty$, then $|\mathbf{E}X| \leqslant \mathbf{E}|X|$.*
(c) *$\mathbf{E}|X| < \infty$ if and only if $D(X) \leqslant \mathbf{E}|X| \leqslant 1 + D(X)$, where*

$$D(X) = \sum_{j=0}^{\infty} j\, \mathbf{P}\{j < |X(\omega)| \leqslant j+1\} = \sum_{j=1}^{\infty} \mathbf{P}\{|X(\omega)| > j\} < \infty$$

(in particular, if $\mathbf{E}|X| < \infty$, then $n\, \mathbf{P}\{|X(\omega)| > n\} \xrightarrow{n \to \infty} 0$).

Proof

(a) If the expected value $\mathbf{E}X = \mathbf{E}X^+ - \mathbf{E}X^-$ is finite, then $\mathbf{E}X^+ < \infty$ and $\mathbf{E}X^- < \infty$. Consequently,

$$\mathbf{E}|X| = \mathbf{E}(X^+ + X^-) = \mathbf{E}X^+ + \mathbf{E}X^- < \infty.$$

Conversely, if $\mathbf{E}|X| < \infty$ then $\mathbf{E}X^+ < \infty$ and $\mathbf{E}X^- < \infty$, then we conclude that $\mathbf{E}X = \mathbf{E}X^+ - \mathbf{E}X^-$ is also finite.

(b) It is easy to see that

$$|\mathbf{E}X| = |\mathbf{E}X^+ - \mathbf{E}X^-| \leq |\mathbf{E}X^+| + |\mathbf{E}X^-| = \mathbf{E}X^+ + \mathbf{E}X^- = \mathbf{E}|X|.$$

(c) We define

$$X_1(\omega) = \sum_{j=0}^{\infty} j \, \mathbf{1}_{\{j < |X(\omega)| \leq j+1\}}, \qquad X_2(\omega) = \sum_{j=0}^{\infty} (j+1) \, \mathbf{1}_{\{j < |X(\omega)| \leq j+1\}}.$$

Evidently, for each $\omega \in \Omega$, we have $X_1(\omega) \leq |X(\omega)| \leq X_2(\omega)$. Note that

$$D(X) = \sum_{j=0}^{\infty} j \, \mathbf{P}\{j < |X(\omega)| \leq j+1\} = \mathbf{E}X_1 \leq \mathbf{E}|X|.$$

On the other hand, we have

$$\mathbf{E}X_2(\omega) = \sum_{j=0}^{\infty} (j+1) \, \mathbf{P}\{j < |X(\omega)| \leq j+1\}$$

$$= \sum_{j=0}^{\infty} j \, \mathbf{P}\{j < |X(\omega)| \leq j+1\} + \sum_{j=0}^{\infty} \mathbf{P}\{j < |X(\omega)| \leq j+1\}$$

$$= D(X) + \mathbf{P}\{|X(\omega)| > 0\} = D(X) + 1 - \mathbf{P}\{X(\omega) = 0\}.$$

This implies that $D(X) < \infty$ if and only if $\mathbf{E}|X| < \infty$. Let us suppose that $D(X) < \infty$. Then we have

$$\infty > D(X) = \sum_{j=0}^{\infty} j \, \mathbf{P}\{j < |X(\omega)| \leq j+1\} = \sum_{j=1}^{\infty} \mathbf{P}\{j < |X(\omega)|\}.$$

Since the latter series is convergent, its partial sums satisfy Cauchy's condition:

$$\sum_{j=1}^{2n} \mathbf{P}\{j < |X(\omega)|\} - \sum_{j=1}^{n} \mathbf{P}\{j < |X(\omega)|\} = \sum_{j=n+1}^{2n} \mathbf{P}\{j < |X(\omega)|\} \xrightarrow{n \to \infty} 0.$$

4.2 General Definition of Expected Value

We also have

$$\sum_{j=n+1}^{2n} \mathbf{P}\{j < |X(\omega)|\} \geqslant \sum_{j=n+1}^{2n} \mathbf{P}\{2n < |X(\omega)|\} = n\,\mathbf{P}\{2n < |X(\omega)|\},$$

which ends the proof of the statement in parentheses.

□

Theorem 4.16 (Lebesgue's Monotone Convergence Theorem) *If a sequence of simple non-negative random variables (X_n), $n = 1, 2, \ldots$, is nondecreasing, then*

$$\mathbf{E}\left(\lim_{n\to\infty} X_n(\omega)\right) = \lim_{n\to\infty} \mathbf{E}\left(X_n(\omega)\right).$$

Proof For every $n \in \mathbb{N}$, we have chosen a nondecreasing sequence of simple random variables $(X_{n,i})$ such that

$$\forall \omega \in \Omega \quad \lim_{i\to\infty} X_{n,i} = X_n.$$

We will use the following notation:

$$X(\omega) = \lim_{n\to\infty} X_n(\omega), \quad Y_n(\omega) = \max\{X_{1,n}(\omega), X_{2,n}(\omega), \ldots, X_{n,n}(\omega)\}.$$

Since for every $i \in \mathbb{N}$ and every $\omega \in \Omega$, the following inequalities hold:

$$X_{n,i}(\omega) \leqslant X_n(\omega) \text{ and } X_1(\omega) \leqslant X_2(\omega) \leqslant \cdots \leqslant X_n(\omega) \leqslant X,$$

we have $Y_n(\omega) \leqslant X_n(\omega) \leqslant X(\omega)$. Hence, we get that

$$\mathbf{E}Y_n \leqslant \mathbf{E}X_n \leqslant \mathbf{E}X.$$

Taking the limit of the second of these two inequalities, we get

$$\lim_{n\to\infty} \mathbf{E}\left(X_n(\omega)\right) \leqslant \mathbf{E}\left(\lim_{n\to\infty} X_n(\omega)\right).$$

To prove the equality, let us note that (Y_n) is a nondecreasing sequence of simple variables that converges to X, which can be easily seen from the following relationship:

$$0 \leqslant X(\omega) - Y_n(\omega) = (X(\omega) - X_n(\omega)) + (X_n(\omega) - Y_n(\omega)),$$

because both components on the right converge to zero. Now, it is enough to calculate $\mathbf{E}X$ from the definition using (Y_n) as the approximating sequence of simple functions. Hence, $\mathbf{E}X = \lim_{n\to\infty} \mathbf{E}Y_n$. □

Theorem 4.17 (Fatou's Lemma) *If (X_n) is a sequence of non-negative random variables, then*

$$\mathbf{E}\left(\liminf_{n\to\infty} X_n\right) \leqslant \liminf_{n\to\infty} \mathbf{E}(X_n).$$

Proof Let $Y_n = \inf\{X_n, X_{n+1}, \ldots\}$. Naturally, $Y_n \leqslant X_n$, thus $\mathbf{E}Y_n \leqslant \mathbf{E}X_n$ and

$$\lim_{n\to\infty} Y_n = \liminf_{n\to\infty} X_n.$$

(Y_n) is a nondecreasing sequence of non-negative variables, hence, by Lebesgue's Monotone Convergence Theorem, we get

$$\mathbf{E}\left(\liminf_{n\to\infty} X_n\right) = \mathbf{E}\left(\lim_{n\to\infty} Y_n\right) = \lim_{n\to\infty} \mathbf{E}(Y_n) \leqslant \liminf_{n\to\infty} \mathbf{E}(X_n),$$

which was to be shown. □

Theorem 4.18 (Lebesgue's Dominated Convergence Theorem) *If $X, Y, X_1, X_2, \ldots$ are random variables such that $\mathbf{E}Y < \infty$, $|X_n| \leqslant Y$ for each $n \in \mathbb{N}$ and $\mathbf{P}\{\omega: X_n(\omega) \to X(\omega)\} = 1$, then $\mathbf{E}|X| < \infty$, $\mathbf{E}X_n \to \mathbf{E}X$ and*

$$\lim_{n\to\infty} \mathbf{E}|X_n - X| = 0.$$

Proof From our assumptions, we get $Y \geqslant 0$. The random variables X_n, $n \in \mathbb{N}$, are integrable because $|X_n| \leqslant Y$, thus $\mathbf{E}|X_n| \leqslant \mathbf{E}Y < \infty$. Since the variables $Y + X_n$ and $Y - X_n$ are non-negative, by Fatou's Lemma we obtain

$$\mathbf{E}\left(\liminf_{n\to\infty}(Y + X_n)\right) \leqslant \liminf_{n\to\infty} \mathbf{E}(Y + X_n)$$

and

$$\mathbf{E}\left(\liminf_{n\to\infty}(Y - X_n)\right) \leqslant \liminf_{n\to\infty} \mathbf{E}(Y - X_n).$$

From the properties of the expected value, the properties of the lower limit and the integrability of the variables, we get

$$\mathbf{E}\left(\liminf_{n\to\infty} X_n\right) \leqslant \liminf_{n\to\infty} \mathbf{E}X_n \leqslant \limsup_{n\to\infty} \mathbf{E}X_n \leqslant \mathbf{E}\left(\limsup_{n\to\infty} X_n\right).$$

By assumption $\liminf_{n\to\infty} X_n = \limsup_{n\to\infty} X_n = X$ a.e., hence, by virtue of the above inequality there exists a limit of the sequence $(\mathbf{E}X_n)_{n\in\mathbb{N}}$ and $\mathbf{E}X = \lim_{n\to\infty} \mathbf{E}X_n$. Obviously, $|X| \leq Y$, thus also $\mathbf{E}|X| \leq \mathbf{E}Y < \infty$.

The proof of the fact that $\mathbf{E}|X_n - X| \to 0$ as $n \to \infty$ is carried out analogously using the obvious inequality $|X_n - X| \leq 2Y$. □

4.2.1 Exercises

229. We say that a random variable X is symmetric if for every $t \in \mathbb{R}$ we have $\mathbf{P}\{X < t\} = \mathbf{P}\{X > -t\}$ (the distributions of X and $-X$ are identical). Suppose that a symmetric variable X has an expected value. Find $\mathbf{E}X$.

230. Let X and Y be random variables. Show that if $\mathbf{E}X$ and $\mathbf{E}Y$ exist then $\mathbf{E}\max\{X, Y\}$ also exists. Does the converse implication hold?

231. Give an example of two random variables X and Y such that $\mathbf{E}X$ and $\mathbf{E}Y$ exist, but $\mathbf{E}(XY)$ does not.

232. Let X be a random variable on $(\Omega, \mathcal{A}, \mathbf{P})$ such that $\mathbf{E}|X| < \infty$. Show that $|X(\omega)| < \infty$ with probability 1.

233. Let X, Y be random variables on $(\Omega, \mathcal{A}, \mathbf{P})$. Prove that if

$$\int_A X(\omega)\mathbf{P}(d\omega) \leq \int_A Y(\omega)\mathbf{P}(d\omega)$$

for any set $A \in \mathcal{A}$, then $X(\omega) \leq Y(\omega)$ with probability 1.

234. Let X and Y be identically distributed random variables. Prove that the following equality does not always hold:

$$\mathbf{E}\frac{X}{X+Y} = \mathbf{E}\frac{Y}{X+Y}.$$

235. Let $X_1, \ldots, X_n$ be independent positive random variables with the same distributions. Prove that for every $k \leq n$

$$\mathbf{E}\frac{X_1 + \cdots + X_k}{X_1 + \cdots + X_n} = \frac{k}{n}.$$

236. Let $\mathbf{E}X = 0$ and $\mathbf{E}|X| = 1$. Find $\mathbf{E}\max\{0, X\}$ and $\mathbf{E}\min\{0, X\}$.

237. What conditions must the numbers a and b meet so that a random variable X having the properties $\mathbf{E}X = a$ and $\mathbf{E}|X| = b$ exists?

238. Let X be a natural-valued random variable with a finite expectation. Prove that

$$\mathbf{E}X = \sum_{k=1}^{\infty} \mathbf{P}\{X \geq k\}.$$

239. Let X and Y be independent random variables with values in the set of natural numbers and let $\mathbf{E}X < \infty$. Prove that

$$\mathbf{E}\min\{X, Y\} = \sum_{k=1}^{\infty} \mathbf{P}\{X \geq k\}\mathbf{P}\{Y \geq k\}.$$

4.3 Functions of Random Variables

Theorem 4.19 *Let $\varphi : \mathbb{R}^n \to \mathbb{R}$ be a Borel function, and $X = (X_1, \ldots, X_n)$ a random vector with distribution $\mathbf{P}_X$. Then:*

$$\mathbf{E}\varphi(X) = \int \cdots \int_{\mathbb{R}^n} \varphi(x)\mathbf{P}_X(\mathrm{d}x)$$

if we know that at least one of the integrals in this formula exists.

Proof If φ is a Borel-measurable function, then, by definition, for every Borel set $B \subset \mathbb{R}$, $\varphi^{-1}(B)$ is also a Borel set, thus

$$\{\varphi(X) \in B\} = \{X \in \varphi^{-1}(B)\} \in \mathcal{F}$$

because X is a random vector. Hence, it follows that $\varphi(X)$ is a random variable. We will show the equality of the integrals mentioned in the theorem in a few steps:

1. If $\varphi = 1_B$ for some Borel set B, then

$$\mathbf{E}\varphi(X) = \mathbf{P}\{X \in B\} = \mathbf{P}_X(B) = \int \cdots \int_B 1\,\mathbf{P}_X(\mathrm{d}x) = \int \cdots \int_{\mathbb{R}^n} \varphi(x)\mathbf{P}_X(\mathrm{d}x).$$

2. If φ is a simple function, i.e., $\varphi = \sum_i x_i 1_{B_i}$ for some Borel sets B_i and $x_i \in \mathbb{R}$, then the desired equality is a result of the linearity of the integral and the equality shown in step 1.
3. If φ is a non-negative Borel function, then it is the limit of a non-decreasing approximating sequence of simple functions. Therefore, it is enough to apply Lebesgue's Monotone Convergence Theorem and the equality obtained in step 2.
4. Let $\varphi = \varphi^+ - \varphi^-$ and assume that $\mathbf{E}\varphi(X)$ exists. Note that $(\varphi(X))^+ = \varphi^+(X)$ and $(\varphi(X))^- = \varphi^-(X)$. Hence, it follows that at least one of the integrals

$$\mathbf{E}\varphi^+(X) \quad \text{or} \quad \mathbf{E}\varphi^-(X)$$

4.3 Functions of Random Variables

is finite. From step 3, we have

$$\mathbf{E}\varphi^+(X) = \int \ldots \int_{\mathbb{R}^n} \varphi^+(x)\mathbf{P}_X(dx), \quad \mathbf{E}\varphi^-(X) = \int \ldots \int_{\mathbb{R}^n} \varphi^-(x)\mathbf{P}_X(dx),$$

which implies the existence of the right-hand integrals and the desired equality. If we assume the existence of the integral $\int \varphi(x)\mathbf{P}_X(dx)$, the reasoning is similar. □

Thanks to this theorem, we can calculate the values of $\mathbf{E}\varphi(X)$ without having to first determine the distribution of $\varphi(X)$ if the distribution of X is already known.

Example 4.20 Assume that $\Omega = [0, 2]$, $\mathbf{P}$ is the normalized Lebesgue measure, i.e., $\mathbf{P}(d\omega) = \frac{1}{2}\mathbf{1}_{[0,2]}(\omega)\,d\omega$, and $\mathcal{F}$ is the σ-field of Borel sets in Ω. We want to calculate $\mathbf{E}X^n$ for the random variable X defined as follows:

$$X(\omega) = \begin{cases} \omega & \text{if } \omega \in [0,1]; \\ 2 - \omega & \text{if } \omega \in (1,2]. \end{cases}$$

Instead of looking for the distribution of the variable X or X^n, let's define a new variable $Y(\omega) = \omega$. It is easy to see that $\mathbf{P}_Y = \mathbf{P}$, so by Theorem 4.19, we get:

$$\mathbf{E}X^n = \mathbf{E}\left(\omega \mathbf{1}_{[0,1]}(Y) + (2-\omega)\mathbf{1}_{(1,2]}(Y)\right)^n$$

$$= \int_{-\infty}^{\infty} \left(x\mathbf{1}_{[0,1]}(x) + (2-x)\mathbf{1}_{(1,2]}(x)\right)^n \frac{1}{2}\mathbf{1}_{[0,2]}(x)\,dx$$

$$= \frac{1}{2}\int_0^1 x^n\,dx + \frac{1}{2}\int_1^2 (2-x)^n\,dx = \frac{1}{n+1}.$$

4.3.1 Exercises

240. Prove that if $\mathbf{E}X^2 < \infty$, then $\mathbf{E}|X| < \infty$.
241. Prove that if $\mathbf{E}X^2 = 0$, then $\mathbf{P}\{X = 0\} = 1$.
242. Let $\Omega = [0, 1]$ and let $\mathbf{P}$ be the Lebesgue measure on Ω. Using Definition 4.7, find the expected values of $X(\omega) = \omega^2$ and $Y(\omega) = \omega^3$.
     Hint: $\sum_{k=1}^{n} k^2 = n(n+1)(2n+1)/6$, $\sum_{k=1}^{n} k^3 = n^2(n+1)^2/4$.
243. Let $\Omega = [0, 3]$ and let $\mathbf{P}$ be normalized Lebesgue measure on Ω. Find the expected value of the following random variables:

$$X(\omega) = \begin{cases} \omega & \text{if } \omega \in [0; 1]; \\ 1 & \text{if } \omega \in (1, 2]; \\ 3 - \omega & \text{if } \omega \in (2, 3]; \end{cases} \quad Y(\omega) = \begin{cases} 2\omega - 1 & \text{if } \omega \in [0; 1]; \\ 2 - \omega & \text{if } \omega \in (1, 2]; \\ 0 & \text{if } \omega \in (2, 3]. \end{cases}$$

244. Prove that if $\mathbf{E}|X|^\alpha < \infty$ for some $\alpha > 0$, then also $\mathbf{E}|X|^\beta < \infty$ for every $\beta \in (0, \alpha)$.
245. Suppose that the expected values of random variables $X_1, \ldots, X_n$ exist and are finite. Show that

$$\mathbf{E}\max\{X_1, \ldots, X_n\} \geqslant \max\{\mathbf{E}X_1, \ldots, \mathbf{E}X_n\};$$
$$\mathbf{E}\min\{X_1, \ldots, X_n\} \leqslant \min\{\mathbf{E}X_1, \ldots, \mathbf{E}X_n\}.$$

4.4 Expected Value for Continuous Type Random Variables

Theorem 4.21 *If the continuous type random variable* $X : \Omega \to \mathbb{R}$ *has density function* $f : \mathbb{R} \to [0, \infty)$ *and* $\mathbf{E}X$ *exists, then*

$$\mathbf{E}X = \int_{-\infty}^{\infty} x f(x)\,\mathrm{d}x.$$

Proof It is enough to prove that the result holds for a non-negative random variable. Let $X \geqslant 0$, which implies the condition $f(x) = 0$ for $x < 0$. By (X_n) we denote the approximating sequence of non-negative simple variables defined as in the proof of Lemma 4.5:

$$X_n = \sum_{k=0}^{n 2^n - 1} \frac{k}{2^n} \mathbf{1}_{I_{k,n}}(X(\omega)) + n \mathbf{1}_{I_n}(X(\omega)),$$

where $I_{k,n} = \left[\frac{k}{2^n}, \frac{k+1}{2^n}\right)$ and $I_n = [n, \infty)$. By definition of the expected value for simple random variables, we have

$$\mathbf{E}X_n = \sum_{k=0}^{n 2^n - 1} \frac{k}{2^n} \mathbf{P}\{X \in I_{k,n}\} + n \mathbf{P}\{X \in I_n\}$$

$$= \sum_{k=0}^{n 2^n - 1} \frac{k}{2^n} \int_{I_{k,n}} f(x)\,\mathrm{d}x + n \int_{I_n} f(x)\,\mathrm{d}x$$

$$= \int_0^\infty \underbrace{\left[\sum_{k=0}^{n 2^n - 1} \frac{k}{2^n} \mathbf{1}_{I_{k,n}}(x) + n \mathbf{1}_{I_n}(x)\right]}_{g_n(x)} f(x)\,\mathrm{d}x.$$

The function $g_n(x)$ approximates the function $g(x) = x$ with a precision of 2^{-n} on the interval $[0, n)$, and outside of this interval, it takes the value of n. Also,

4.4 Expected Value for Continuous Type Random Variables

$g_n(x) \nearrow g(x)$ as $n \to \infty$, for every $x \in \mathbb{R}$. The existence of the limit of the integrals $\int g_n(x) f(x) \, dx$ as $n \to \infty$ is due to the existence of $\mathbf{E}X$. Hence, when we go to the limit, we get

$$\mathbf{E}X = \lim_{n \to \infty} \mathbf{E}X_n = \lim_{n \to \infty} \int_0^\infty g_n(x) f(x) \, dx$$
$$= \int_0^\infty \left(\lim_{n \to \infty} g_n(x) \right) f(x) \, dx = \int_0^\infty x f(x) \, dx,$$

which was to be shown. □

Example 4.22 To calculate the expected value of a variable X with distribution $\Gamma(b, p)$ (density $\frac{b^p}{\Gamma(p)} x^{p-1} e^{-bx} \mathbf{1}_{(0,\infty)}(x)$), let us recall first that for any $b > 0$, $p > 0$

$$\int_0^\infty x^{p-1} e^{-bx} \, dx = \frac{\Gamma(p)}{b^p}.$$

Hence, we have

$$\mathbf{E}X = \int_0^\infty x \frac{b^p}{\Gamma(p)} x^{p-1} e^{-bx} \, dx = \frac{b^p}{\Gamma(p)} \int_0^\infty x^p e^{-bx} \, dx = \frac{b^p}{\Gamma(p)} \frac{\Gamma(p+1)}{b^{p+1}}.$$

Since $\Gamma(p+1) = p \, \Gamma(p)$, we finally get $\mathbf{E}X = p/b$.

Theorem 4.23 *If a random variable X has a continuous type distribution with density $f(x)$, and φ is a Borel function, then*

$$\mathbf{E}\varphi(X) = \int_{-\infty}^\infty \varphi(x) f(x) \, dx$$

if at least one of the integrals in this formula exists.

Proof If $\varphi(x) = \mathbf{1}_B(x)$, then

$$\mathbf{E}\varphi(X) = \mathbf{P}\{X \in B\} = \int_B f(x) \, dx = \int_{-\infty}^\infty \varphi(x) f(x) \, dx.$$

Now, it suffices to repeat the standard reasoning already used: show that equality holds for simple functions φ, by Lebesgue's Monotone Convergence Theorem extend the equality to non-negative Borel functions, then consider positive and negative parts separately for any Borel function φ. □

4.4.1 Exercises

246. Calculate the expected value for the following distributions: a) uniform on the interval $[a, b]$; b) exponential with parameter λ; c) beta distribution with parameters $p, q > 0$.
247. Calculate $\mathbf{E}X$ and $\mathbf{E}X^2$ for a random variable X with the normal distribution $N(m, \sigma)$.
248. Does there exist an expected value for a Cauchy distributed random variable with parameters $a > 0$ and $m \in \mathbb{R}$?
249. Calculate $\mathbf{E}X^n$ for a variable X having the following distributions: a) gamma with parameters $b, p > 0$; b) beta with parameters $p, q > 0$.
250. Margaret has a standard A4 (210mm × 297mm) sheet of paper and she folds it perpendicularly to the longer side at a randomly chosen place. What is the expected value of the area of the larger rectangle obtained after folding the sheet?

4.5 Expected Value as a Lebesgue–Stieltjes Integral

One more approach to calculating the expected value is to use the distribution of the variable in question, specifically, its distribution function.

Definition 4.24 Assume that F is the distribution function of the probability measure $\mathbf{P}$ on Ω and let $g : \mathbb{R} \to \mathbb{R}$ be a Borel measurable function. The *Lebesgue–Stieltjes integral* of the function g with respect to increments of the distribution function F is defined by the formula:

$$\int_{-\infty}^{\infty} g(x)\,dF(x) \stackrel{\text{def}}{=} \int_{-\infty}^{\infty} g(x)\mathbf{P}(dx) = \mathbf{E}g(X(\omega)),$$

where $X(\omega)$ is a random variable with distribution $\mathbf{P}$.

Note that this is not a new kind of integral for us. We already know that any left-continuous nondecreasing function F such that

$$\lim_{t \to -\infty} F(t) = 0, \qquad \lim_{t \to \infty} F(t) = 1$$

is the distribution function of some probability distribution $\mathbf{P}$, which uniquely defines the right side of the formula. If the function g is uniformly continuous on $\mathbb{R}$, then the Lebesgue–Stieltjes integral is sometimes referred to as the *Riemann–Stieltjes integral* or simply the *Stieltjes integral*. These distinctions are not important in the context of the general theory of measure, and the integral $\int g(x)\,dF(x)$ is best treated only as another way of writing the integrals $\int g(x)\mathbf{P}(dx)$ or $\mathbf{E}g(X)$.

4.5 Expected Value as a Lebesgue–Stieltjes Integral

However, this notation turns out to be very helpful in calculating the expected values for distributions that are convex combinations of a discrete distribution and an absolutely continuous distribution. Therefore, let us assume that the distribution function $F(x)$ of the variable X has at most a countable number of jumps at points x_i of heights p_i, while away from the jump points it is differentiable and

$$\int_{-\infty}^{\infty} F'(x)\,dx + \sum_i p_i = 1.$$

Then, writing the expected value as an integral with respect to increments of the distribution function seems only natural:

$$EX = \int_{-\infty}^{\infty} x\,dF(x) = \sum_i x_i p_i + \int_{-\infty}^{\infty} x F'(x)\,dx.$$

Such a distribution is a convex combination of a discrete and an absolutely continuous distribution. To see this, let $f(x) = F'(x)$ and define

$$\alpha = \sum_i p_i, \quad \mathbf{P}_1(dx) = \alpha^{-1} \sum_i p_i \delta_{x_i}(dx), \quad \mathbf{P}_2(dx) = (1-\alpha)^{-1} f(x)\,dx.$$

If $\mathbf{P}$ is a distribution with distribution function F, then

$$\mathbf{P} = \alpha \mathbf{P}_1 + (1-\alpha)\mathbf{P}_2.$$

It is not difficult to verify that $\mathbf{P}_1$ and $\mathbf{P}_2$ are probability distributions, $\mathbf{P}_1$ is a discrete distribution and P_2 is absolutely continuous.

In the last exercise of this chapter, we will see that not all probability distributions are convex combinations of an absolutely continuous distribution and a discrete distribution.

4.5.1 Exercises

251. Let X be a positive random variable with distribution function F such that $EX^\alpha < \infty$ for some $\alpha > 0$. Show that for any $c > 0$, we have

$$EX^\alpha = \alpha \int_0^\infty x^{\alpha-1}(1 - F(x))\,dx,$$

$$\mathbf{E}(\min\{X^\alpha, c\}) = \alpha \int_0^c x^{\alpha-1}(1 - F(x))\,dx.$$

252. Let X and Y be positive random variables with distribution functions $F(x)$ and $G(x)$. Prove that if $F(x) \leq G(x)$ for every $x \in \mathbb{R}$, then $\mathbf{E}X \geq \mathbf{E}Y$.

253. Let $0 < 2a < 1$. Find the expected value of a random variable with the following distribution functions:

(a) $F(x) = \begin{cases} e^x & \text{if } x \leq -1; \\ 0.5 & \text{if } -1 < x \leq 1; \\ 1 - \frac{1}{3x^3} & \text{if } x > 1; \end{cases}$ (b) $F(x) = \begin{cases} 0 & \text{if } x \leq 0; \\ x + a & \text{if } 0 < x \leq \frac{1}{2}; \\ 1 & \text{if } x > \frac{1}{2}. \end{cases}$

254. Let us build the Cantor set on the interval $[0, 1]$. In the first step ($n = 1$), we divide it into three equal intervals and denote the middle (open) interval by $I_{1,1}$. In the second step ($n = 2$), the remaining segments $[0, \frac{1}{3}]$ and $[\frac{2}{3}, 1]$ are divided into three equal parts and the middle ones are denoted by $I_{1,2}, I_{2,2}$. Continuing this procedure, we get a sequence of open intervals $I_{k,n}$, $n \in \mathbb{N}$, $k = 1, 2, \ldots, 2^{n-1}$. Prove that the Cantor set defined by the following formula is non-empty:

$$C = [0, 1] \setminus \bigcup_{n \in \mathbb{N}} \bigcup_{k=1}^{2^{n-1}} I_{k,n}.$$

255. (continued) Now, we are able to build Cantor's function, also called the Devil's Staircase, $F : \mathbb{R} \to [0, 1]$ as follows: $F(x) = 0$ for $x \leq 0$, $F(x) = 1$ for $x > 1$, and

$$F(x) = \begin{cases} 2^{-n}(2k - 1) & \text{if } x \in I_{k,n}; \\ \sup_{t < x, t \notin C} F(t) & \text{if } x \in C. \end{cases}$$

(a) Sketch the graph of F on the set $I_{1,1} \cup I_{1,2} \cup I_{2,2} \cup I_{1,3} \cup I_{2,3} \cup I_{3,3} \cup I_{4,3}$.
(b) Prove that F is a distribution function.
(c) Prove that F is a continuous function, hence, the random variable with distribution function F is not discrete and has no atoms.
(d) Prove that for every $x \notin C$ the derivative $F'(x)$ exists and $F'(x) = 0$, hence the random variable with distribution function F is not of continuous type.

Chapter 5
Random Variable Parameters

5.1 Quantiles, Median, Moments, Variance, Skewness and Kurtosis

Often, especially in mathematical statistics, numerical parameters are crucial in describing a random variable X. In this chapter, we describe the most important of them.

- A *quantile of order* $p \in (0, 1)$ of a random variable X, or its distribution, is a number α_p for which the following inequalities hold:

$$\mathbf{P}\{X \leqslant \alpha_p\} \geqslant p, \quad \mathbf{P}\{X \geqslant \alpha_p\} \geqslant 1 - p.$$

The quantile of $p = \frac{1}{2}$ is called the *median*. Of course, for symmetric distributions, $\alpha_{1/2} = 0$. Sometimes, in statistics, quantiles are referred to as *percentiles* if the parameter p is a probability expressed as a percentage.

Note that the quantile α_p is not necessarily uniquely determined. If the distribution of X is given by density:

$$f(x) = \frac{2}{3}x^{-3}\mathbf{1}_{(-\infty,-1)}(x) + \frac{2}{3}x^{-2}\mathbf{1}_{(1,\infty)}(x),$$

then the $\alpha_{1/3}$ quantile can be any number from the interval $[-1, 1]$.

The second class of numerical parameters of the variable X are its moments: ordinary, central, and absolute. When defining moments, we will assume that the relevant integrals exist and are finite.

- The *raw or crude moment*, or simply the *moment* of order $k \in \mathbb{N}$ of the variable X, is the expected value of the variable X^k:

$$m_k = \mathbf{E}X^k.$$

- The *absolute moment* of order $r > 0$ is the expected value of the variable $|X|^r$:

$$\beta_r = \mathbf{E}|X|^r.$$

- The *central moment* of order $k \in \mathbb{N}$ is the expected value of the random variable $(X - \mathbf{E}X)^k$:

$$\mu_k = \mathbf{E}(X - \mathbf{E}X)^k.$$

- The *variance* of the variable X is the second central moment of this variable:

$$\text{Var}\, X = \mu_2 = \mathbf{E}(X - \mathbf{E}X)^2.$$

Note that from Newton's binomial formula, one can easily obtain the connection between the raw and central moments:

$$\mu_k = \mathbf{E}(X - \mathbf{E}X)^k = \sum_{j=0}^{k} \binom{k}{j}(-1)^{k-j} \mathbf{E}X^j \, (\mathbf{E}X)^{k-j}$$

$$= \sum_{j=0}^{k} \binom{k}{j}(-1)^{k-j} m_j m_1^{k-j}.$$

In particular, for $k = 2$, we have

$$\text{Var}\, X = m_2 - m_1^2 = \mathbf{E}X^2 - (\mathbf{E}X)^2.$$

The variance increases as the variable deviates more from its expected value. The variance is zero if and only if the random variable is constant almost everywhere, because $\mathbf{E}(X - \mathbf{E}X)^2 = 0$ if and only if $X = \mathbf{E}X$ with probability one. Moreover, it is easy to see that:

- $\text{Var}(cX) = c^2 \text{Var}(X)$;
- $\text{Var}(X + c) = \text{Var}(X)$;
- if X and Y are independent then $\text{Var}(X + Y) = \text{Var}(X) + \text{Var}(X)$.

- The square root of the variance is called the *standard deviation* and is denoted by the symbol σ_X:

$$\sigma_X = \sqrt{\text{Var}(X)}.$$

We say that the random variable X is standardized if $\text{Var}\, X = 1$ or if we consider the random variable X/σ_X instead of X.

5.1 Quantiles, Median, Moments, Variance, Skewness and Kurtosis

- The third standardized moment ν_3 of a random variable X is often called its *skewness* and is defined as:

$$\nu_3 = \frac{E(X - EX)^3}{\sigma_X^3}.$$

Skewness is a measure of the asymmetry of the probability distribution of a real-valued random variable about its mean. The skewness value can be positive, zero, negative, or undefined.

- The fourth standardized moment ν_4 of a random variable X is called its *kurtosis* (from Greek: kurtos, meaning "curved"):

$$\nu_4 = \frac{E(X - EX)^4}{\sigma_X^4}.$$

Kurtosis is a measure of the thickness of the distribution's tail for the real-valued random variable X.

Example 5.1 A random variable X with distribution $\Gamma(p, b)$ has all moments of positive order and some of negative order because for $p + r > 0$, we have

$$EX^r = \int_0^\infty x^r \frac{b^p}{\Gamma(p)} x^{p-1} e^{-bx} \, dx = \frac{b^p}{\Gamma(p)} \int_0^\infty x^{p+r-1} e^{-bx} \, dx$$

$$= \frac{b^p}{\Gamma(p)} \cdot \frac{\Gamma(p+r)}{b^{p+r}} = \frac{\Gamma(p+r)}{b^r \Gamma(p)} < \infty.$$

Example 5.2 We say that a random variable X has a *Pareto distribution* with parameter $\alpha > 0$ if its density is given by the formula:

$$f(x) = \begin{cases} \alpha x^{-\alpha-1} & \text{if } x > 1; \\ 0 & \text{if } x \leqslant 1. \end{cases}$$

This variable has all moments of order r (including negative) as long as $r < \alpha$, because then

$$EX^r = \int_1^\infty x^r \alpha x^{-\alpha-1} \, dx = \frac{\alpha}{r - \alpha} < \infty.$$

The next three theorems describe other connections between the moments of random variables.

Theorem 5.3 (Schwarz's Inequality) *If* $EX^2 < \infty$ *and* $EY^2 < \infty$, *then*

$$|E(XY)| \leqslant \sqrt{EX^2 EY^2}.$$

Proof For any real number t, the following inequality is true:

$$\mathbf{E}(tX+Y)^2 = t^2\mathbf{E}X^2 + 2t\mathbf{E}(XY) + \mathbf{E}Y^2 \geqslant 0.$$

If we treat the left-hand side of the above inequality as a quadratic function of t, it becomes obvious that $\Delta \leqslant 0$. Hence,

$$\Delta = 4\left(\mathbf{E}(XY)\right)^2 - 4\mathbf{E}X^2\mathbf{E}Y^2 \leqslant 0,$$

which ends the proof. □

Theorem 5.4 (Jensen's Inequality) *Let us assume that* $\mathbf{E}|X| < \infty$. *If g is a convex function on $\mathbb{R}$, then*

$$g(\mathbf{E}X) \leqslant \mathbf{E}g(X).$$

Proof If g is a convex function, then for every point x_0, there exists a constant $m(x_0)$ such that

$$g(x) \geqslant g(x_0) + m(x_0)(x - x_0).$$

The constant $m(x_0)$ is the slope of the line supporting the graph of the function g at the point x_0. When substituting $X \to x$ and $\mathbf{E}X \to x_0$ and taking the expected value of both sides of this inequality, we get the result. □

Theorem 5.5 (Hölder's Inequality) *Assume that $p, q > 1$ satisfy the condition $\frac{1}{p} + \frac{1}{q} = 1$. If $\mathbf{E}|X|^p < \infty$ and $\mathbf{E}|Y|^q < \infty$, then $\mathbf{E}|XY| < \infty$ and*

$$\mathbf{E}|XY| \leqslant \left(\mathbf{E}|X|^p\right)^{1/p} \left(\mathbf{E}|Y|^q\right)^{1/q}.$$

Proof Note that the function $\log x$ is concave on the half-line $(0, \infty)$. It follows that for any $\alpha, \beta > 0$, $\alpha + \beta = 1$ and any $x, y > 0$, the following inequality holds:

$$\log(\alpha x + \beta y) \geqslant \alpha \log x + \beta \log y = \log(x^\alpha y^\beta).$$

Hence,

$$\alpha x + \beta y \geqslant x^\alpha y^\beta.$$

Now, it is enough to substitute

$$\alpha = \frac{1}{p}, \quad \beta = \frac{1}{q}, \quad x^{1/p} = \frac{|X|}{(\mathbf{E}|X|^p)^{1/p}}, \quad y^{1/p} = \frac{|Y|}{(\mathbf{E}|Y|^q)^{1/q}}$$

and calculate the expected value of both sides of this inequality. □

5.1.1 Exercises

256. Find the quantiles of orders 0.25; 0.5 and 0.75 for the uniform distribution on the interval $[a, b]$.
257. Calculate the median and the quantile of order 0.25 for the following distributions: (a) exponential; (b) Cauchy.
258. Calculate all moments, skewness and kurtosis of a random variable X with the beta distribution with parameters $p, q > 0$.
259. Calculate the four first moments, skewness and kurtosis for the random variable X describing the result of rolling one die.
260. Calculate the expected values and variances for the following discrete distributions: (a) Bernoulli; (b) geometric; (c) Poisson; (d) Pascal.
261. Calculate the variance for the following absolutely continuous distributions: (a) uniform on the interval $[a, b]$; (b) exponential; (c) gamma; (d) beta.
262. Calculate the expectation, variance, skewness and kurtosis for the variable $Z = XY$ if these parameters are known for the independent variables X and Y.
263. Let $\Omega = [0, 1]$ and let $\mathbf{P}$ be the Lebesgue measure on Ω. Find the expected value and the variance of the following random variables:

$$X(\omega) = \omega - 1/2; \quad Y(\omega) = (\omega - 1/2)^2; \quad Z(\omega) = \sin \pi \omega; \quad W(\omega) = \sin 2\pi \omega.$$

264. Prove that if $\mathbf{P}\{0 < X < 1\} = 1$ for a random variable X, then $\operatorname{Var} X < \mathbf{E} X$.
265. Random variables X and Y are independent and such that $\mathbf{E} X = 1$, $\mathbf{E} Y = 2$, $\operatorname{Var} X = 1$ and $\operatorname{Var} Y = 4$. Find the expected values of the following random variables: (a) $Z = X^2 + 2Y^2 - XY - 4X + Y + 4$; (b) $W = (X + Y + 1)^2$.
266. Prove that if $\operatorname{Var} X = 0$, then there exists an $a \in \mathbb{R}$ such that $\mathbf{P}\{X = a\} = 1$.
267. Prove that

$$\operatorname{Var} X = \inf_{a \in \mathbb{R}} \mathbf{E} (X - a)^2.$$

268. Show that for every random variable with a finite first moment and median m_e, we have

$$\inf_{a \in \mathbb{R}} \mathbf{E}|X - a| = \mathbf{E}|X - m_e|.$$

269. Let X and Y be independent random variables. Prove that

$$\operatorname{Var}(XY) \geqslant \operatorname{Var}(X)\operatorname{Var}(Y).$$

What conditions must X and Y satisfy for equality to occur in this inequality?
270. Suppose a random variable X has a symmetric distribution. Prove that for any real number a the following inequality holds:

$$\mathbf{E}|X + a| \geqslant \mathbf{E} X.$$

271. A random variable X satisfies the condition $\mathbf{E}|X|^\alpha < \infty$ for some $\alpha > 0$. Show that for every $c \in \mathbb{R}$

$$\mathbf{E}|X - c|^\alpha < \infty.$$

Hint. $(a + b)^\alpha \leqslant a^\alpha + b^\alpha$ for $\alpha \in (0, 1]$ and $a, b > 0$, while for $\alpha > 1$ and $a, b > 0$, we have $(a + b)^\alpha \leqslant 2^{\alpha-1}(a^\alpha + b^\alpha)$.

272. Let us assume that all the moments of a random variable X are finite. Show that $\psi(u) = \log(\mathbf{E}|X|^u)$ is a convex function for $u \geqslant 0$. What conditions must the variable satisfy for ψ to be a linear function?

273. **Minkowski's inequality.** Prove that for any $a \geqslant 1$:

$$\left(\mathbf{E}|X + Y|^a\right)^{1/a} \leqslant \left(\mathbf{E}|X|^a\right)^{1/a} + \left(\mathbf{E}|Y|^a\right)^{1/a}.$$

Hint. For $a = 1$, the inequality is obvious. For $a > 1$, we have

$$\mathbf{E}|X + Y|^a \leqslant \mathbf{E}\big(|X|\,|X + Y|^{a-1}\big) + \mathbf{E}\big(|Y|\,|X + Y|^{a-1}\big).$$

Now, it is enough to apply Hölder's inequality twice.

5.2 Chebyshev's Inequality

The significance of the moments of a random variable is well described by the following theorem and its generalizations:

Theorem 5.6 (Chebyshev's Inequality) *If X is a random variable and* $\operatorname{Var} X < \infty$, *then for any $\varepsilon > 0$, the following inequality holds:*

$$\mathbf{P}\{\omega : |X - \mathbf{E}X| > \varepsilon\} \leqslant \frac{\operatorname{Var} X}{\varepsilon^2}.$$

If X is a non-negative random variable and $\mathbf{E}X < \infty$, *then for any $\varepsilon > 0$*

$$\mathbf{P}\{\omega : X > \varepsilon\} \leqslant \frac{\mathbf{E}X}{\varepsilon}.$$

Proof Of course, the finite variance assumption in the first condition and the finite expected value assumption in the second are not relevant. However, the estimates obtained without these assumptions do not seem particularly interesting.

Let us assume that $\operatorname{Var} X < \infty$ and let $\varepsilon > 0$. If

$$A = \{\omega : |X(\omega) - \mathbf{E}X| > \varepsilon\} \text{ and } A' = \Omega \setminus A,$$

5.2 Chebyshev's Inequality

then

$$\text{Var} X = \int_A |X - \mathbf{E}X|^2 \, d\mathbf{P} + \int_{A'} |X - \mathbf{E}X|^2 \, d\mathbf{P}$$
$$\geq \int_A |X - \mathbf{E}X|^2 \, d\mathbf{P} \geq \varepsilon^2 \int_A d\mathbf{P}$$
$$= \varepsilon^2 \mathbf{P}(A) = \varepsilon^2 \mathbf{P}\{\omega : |X - \mathbf{E}X| > \varepsilon\},$$

which ends the proof of the first part of the theorem. The proof of the second part is analogous. □

Chebyshev's inequality and a whole series of similar results can be obtained as corollaries of the following, slightly more general, theorem.

Theorem 5.7 *Suppose X is a random variable and g is a non-negative even function defined on $\mathbb{R}$ that is nondecreasing on $[0, \infty)$. Then, for every $\varepsilon > 0$,*

$$\mathbf{P}\{\omega : |X(\omega)| \geq \varepsilon\} \leq \mathbf{E}[g(X)]/g(\varepsilon).$$

Proof If $g \geq 0$, then the integral $\mathbf{E}g(X)$ exists although it may be equal to $+\infty$. Let

$$A = \{\omega : |X(\omega)| \geq \varepsilon\}.$$

On the set A, we have $g(X) \geq g(\varepsilon)$. Therefore, following the proof of Chebyshev's inequality, we get

$$\mathbf{E}g(X) = \int_A g(X) \, d\mathbf{P} + \int_{A'} g(X) \, d\mathbf{P}$$
$$\geq \int_A g(X) \, d\mathbf{P} \geq g(\varepsilon) \int_A d\mathbf{P}$$
$$= g(\varepsilon) \mathbf{P}(A) = g(\varepsilon) \mathbf{P}\{\omega : |X| > \varepsilon\},$$

which was to be shown. □

Inequalities of the Chebyshev type turn out to be very useful when we want to estimate the probability value of certain events, especially when exact calculations are particularly laborious.

Example 5.8 A game involves tossing a symmetrical coin 160 times. We want to determine the interval I into which the number of heads obtained will fall with a probability of at least 0.9.

Let X be the number of heads obtained. We already know that $p = \frac{1}{2}$, $\mathbf{E}X = 80$, $\text{Var} X = 40$. This means that the most probable events occur around 80 and it can be assumed that the optimal interval will be symmetric around this number. From

Chebyshev's inequality, we get

$$\mathbf{P}\{|X - 80| < \varepsilon\} = 1 - \mathbf{P}\{|X - 80| \geqslant \varepsilon\} \geqslant 1 - \frac{40}{\varepsilon^2}.$$

The conditions will be met if we select ε such that $1 - 40\varepsilon^{-2} \geqslant 0.9$, which leads to the condition $\varepsilon \geqslant 20$. We want to choose the interval to be as narrow as possible, therefore we take $\varepsilon = 20$, and we get:

$$0.9 \leqslant \mathbf{P}\{|X - 80| < \varepsilon\} = \mathbf{P}\{X \in (60, 100)\}.$$

The random variable X is discrete and can only take natural values. We can use this information to improve the estimate and write:

$$0.9 \leqslant \mathbf{P}\{X \in [61, 99]\}.$$

5.2.1 Exercises

274. We roll a single die 180 times. Estimate with a probability of 0.9 the number of 6's obtained.
275. Using Chebyshev's inequality, it was calculated that the probability that the number of heads in a series of symmetric coin tosses will differ from its expected value by more than 25% of that expected value is not greater than $1/160$. At least how many tosses did this series consist of?
276. Prove the *three-sigma rule*: if $\mathrm{Var}X = \sigma^2 < \infty$, then

$$\mathbf{P}\{\omega : |X - \mathbf{E}X| < 3\sigma\} \geqslant \frac{8}{9} = 0.888888\ldots$$

277. A random variable X has the distribution $N(m, \sigma)$. Compare the estimate of $\mathbf{P}\{|X - m| < 3\sigma\}$ obtained in the previous exercise with its real value.
278. Let $f : \mathbb{R} \to \mathbb{R}$ be an even function that is measurable and is non-decreasing on the positive half-line. Show that for any random variable X satisfying the condition $|X(\omega)| < c$ for each $\omega \in \Omega$, the following property holds:

$$\forall \varepsilon > 0 \quad \frac{\mathbf{E}f(X) - f(\varepsilon)}{f(c)} \leqslant \mathbf{P}\{|X - \mathbf{E}X| \geqslant \varepsilon\} \leqslant \frac{\mathbf{E}f(X - \mathbf{E}X)}{f(\varepsilon)}.$$

279. **Markov's inequality.** Prove that for any random variable X, any $p > 0$ and $t > 0$, the following inequality holds:

$$\mathbf{P}\{|X| > t\} \leqslant t^{-p}\mathbf{E}|X|^p.$$

280. Prove that for any random variable X, any $p > 0$ and $t > 0$, the following inequality holds:
$$\mathbf{P}\{|X| > t\} \leqslant e^{-pt} \mathbf{E}e^{pX}.$$

5.3 Parameters of Random Vectors

Definition 5.9 The *covariance* of random variables X and Y has the following value:
$$\mathrm{Cov}(X, Y) = \mathbf{E}\Big((X - \mathbf{E}X)(Y - \mathbf{E}Y)\Big).$$

It is easy to see that $\mathrm{Cov}(X, Y)$ can also be calculated from the following formula:
$$\mathrm{Cov}(X, Y) = \mathbf{E}(XY) - \mathbf{E}X \cdot \mathbf{E}Y.$$

The covariance is a parameter that to some extent describes the degree of dependence of random variables, although the correlation factor is more commonly used as a measure of dependence. It is defined as the covariance of standardized variables:

Definition 5.10 If the random variables X and Y have finite and non-zero variances, then their *correlation factor* is given by
$$\varrho(X, Y) = \frac{\mathrm{Cov}(X, Y)}{\sqrt{\mathrm{Var}X \, \mathrm{Var}Y}}.$$

If $\varrho(X, Y) = 0$, we say that the variables X and Y are *uncorrelated*. If $\varrho(X, Y) > 0$, then we say that the variables are *positively correlated*, and if $\varrho(X, Y) < 0$, they are *negatively correlated*.

Remark 5.11 It is easy to check that independent random variables are uncorrelated. It is not true, however, that if the variables are uncorrelated, they are independent!

Theorem 5.12 *For any random variables X, Y with finite second moments, the following property holds:*
$$-1 \leqslant \varrho(X, Y) \leqslant 1.$$

Moreover, $|\varrho(X, Y)| = 1$ if and only if there exist constants $a, b, c \in \mathbb{R}$, $ab \neq 0$, such that
$$\mathbf{P}\Big\{\omega : aX(\omega) + bY(\omega) + c = 0\Big\} = 1.$$

Proof The first part of the theorem follows from the Schwarz inequality applied to the variables $(X - EX)$ and $(Y - EY)$:

$$|E(X - EX)(Y - EY)| \leq \sqrt{E(X - EX)^2 E(Y - EY)^2} = \sqrt{\text{Var} X \text{Var} Y}.$$

It is easy to see that if $Y = aX + b$ with probability 1, then $\varrho(X, Y) = 0$. Let us assume that $\varrho(X, Y) = 1$ and let

$$U = \frac{X - EX}{\sqrt{\text{Var} X}} - \frac{Y - EY}{\sqrt{\text{Var} Y}}.$$

Note that $\text{Var} U = 2 - 2\varrho(X, Y) = 0$. Now, it is enough to apply the already known fact that only the variance of a constant can be equal to zero. If $\varrho(X, Y) = -1$, as an auxiliary variable we should take

$$U = \frac{X - EX}{\sqrt{\text{Var} X}} + \frac{Y - EY}{\sqrt{\text{Var} Y}}.$$

□

Definition 5.13 The *expected value of a random vector* $X = (X_1, \ldots, X_n)$ is the vector of the expected values of its components $EX = (EX_1, \ldots, EX_n)$. The *covariance matrix* of a random vector X is the matrix $\Sigma = (\sigma_{ij})_{i,j=1}^n$, where

$$\sigma_{ij} = \text{Cov}(X_i, X_j).$$

Recall that a square matrix $\Sigma = (\sigma_{ij})$ of dimension $n \times n$ is positive definite (non-negative definite) if $\sum_{i,j=1}^n \sigma_{ij} t_i t_j > 0$ ($\sum_{i,j=1}^n \sigma_{ij} t_i t_j \geq 0$) for any $t_1, \ldots, t_n \in \mathbb{R}$.

Lemma 5.14 *The covariance matrix of any random vector* $X = (X_1, \ldots, X_n)$ *is a non-negative definite matrix. If the random variables* $(X_i - EX_i)$ *are linearly independent, then the covariance matrix of the vector* X *is positive definite.*

Proof The covariance matrix for the vector X is equal to $\left(\text{Cov}(X_i, X_j)\right)_{i,j}$. Hence,

$$\sum_{i,j=1}^n t_i t_j \text{Cov}(X_i, X_j) = \sum_{i,j=1}^n t_i t_j E\left((X_i - EX_i)(X_j - EX_j)\right)$$

$$= E\left(\sum_{i=1}^n t_i (X_i - EX_i)\right)^2 \geq 0.$$

In the above inequality, equality holds only if there exist constants $t_1, \ldots, t_n$ such that $\sum_{i=1}^n t_i (X_i - EX_i) = 0$ with probability 1, in other words, when the variables $X_i - EX_i$, $i = 1, \ldots, n$ are linearly dependent. □

5.3.1 Copulas

In stochastic modeling of real phenomena, it is often necessary to construct a random vector (X_1, X_2) with given parameters $m_1 = \mathbf{E}X_1$, $m_2 = \mathbf{E}X_2$, $\sigma_1^2 = \text{Var}X_1$, $\sigma_2^2 = \text{Var}X_2$ and $\varrho = \varrho(X_1, X_2)$. We already know (see Theorem 3.17) that on the basis of a random variable X with a uniform distribution on the interval $[0, 1]$, it is possible to construct a variable X_1 with arbitrarily chosen distribution function F by substituting $X_1 = F(X)$. Thus, the crux of the problem is the construction of a random vector (X, Y) whose marginal distributions are uniform on the intervals $[0, 1]$, and the correlation coefficient takes a predetermined value of ϱ. Two-dimensional distributions with such properties are called *copulas*. Here, we present one of the simplest examples of a copula.

Consider a random vector (X, Y) uniformly distributed on the frame shown in Fig. 5.1. The length of the entire frame is $2\sqrt{2}$. It is not difficult to see that the part of the frame on the left side of the line $x = t$, $t \in [0, 1]$, has length $2\sqrt{2}\,t$. This means that the F_X marginal distribution function of the variable X has the following form

$$F_X(t) = \begin{cases} 0 & \text{if } t \leq 0; \\ t & \text{if } 0 < t \leq 1; \\ 1 & \text{if } t \geq 1. \end{cases}$$

We see that F_X is the distribution function of the uniform distribution on the interval $[0, 1]$. Due to the symmetry, the same can be said about the distribution of the random variable Y. It follows that

$$\mathbf{E}X = \mathbf{E}Y = \frac{1}{2}, \quad \text{Var}X = \text{Var}Y = \frac{1}{12}.$$

To calculate $\mathbf{E}(XY)$, we need to integrate the product of xy over the curve we call a frame. We will do this separately for each interval included in the frame, remembering that the density along this curve is constant and equal to $2^{-3/2}$, hence,

Fig. 5.1 A frame on which a random vector (X, Y) is uniformly distributed

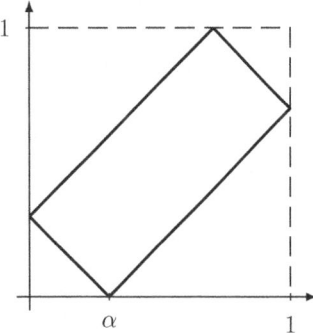

the frame length increment dl is equal to $\sqrt{2}\,dx$. We obtain

$$\mathbf{E}(XY) = \frac{1}{2\sqrt{2}} \left\{ \int_0^\alpha x(\alpha - x)\sqrt{2}\,dx + \int_\alpha^1 x(x - \alpha)\sqrt{2}\,dx \right.$$
$$\left. + \int_0^{1-\alpha} x(x + \alpha)\sqrt{2}\,dx + \int_{1-\alpha}^1 x(2 - x - \alpha)\sqrt{2}\,dx \right\}$$
$$= \frac{1}{6}\left(2\alpha^3 - 3\alpha^2 + 2\right).$$

Hence, $\mathrm{Cov}(X, Y) = (4\alpha^3 - 6\alpha^2 + 1)/12$ and

$$\varrho(X, Y) = \varrho(\alpha) = 4\alpha^3 - 6\alpha^2 + 1.$$

The function $\varrho(\alpha)$ is a continuous function of the argument α. Moreover, $\varrho(0) = 1$, $\varrho(1) = -1$. It follows that for any predetermined value of $\varrho \in [-1, 1]$, we can find $\alpha \in [0, 1]$ such that $\varrho(\alpha) = \varrho$, which was to be shown.

5.3.2 H. Markowitz's Investing Theory

In 1990, H. Markowitz was awarded the Nobel Prize in Economics for his work on investing methods in the stock exchange. One of the methods developed by him was based on the observation that if the shares of two different companies are stochastically negatively correlated, then a possible decrease in the value of one of them should coincide with an increase in the value of the other. If we choose both stocks from among those that tend to increase, we can only earn!

Unfortunately, like everything that concerns processes that cannot be described in a deterministic way, there is a risk that we will suffer a loss. Therefore, a portfolio of shares should be constructed in such a way that this risk is minimal.

We start by observing the quotations of several or a dozen companies on the stock exchange for a long period of time. On this basis, using statistical methods, we can select two types of shares A and B with an upward trend and a negative correlation. Let X_1, X_2 be random variables which describe the prices of the stocks A and B, respectively. We statistically determine the following parameters: $m_1 = \mathbf{E}X_1$, $m_2 = \mathbf{E}X_2$, $\sigma_1^2 = \mathrm{Var}X_1$, $\sigma_2^2 = \mathrm{Var}X_2$ and $\varrho = \varrho(X_1, X_2)$. By assumption, we choose X_1, X_2 with $\varrho < 0$, preferably with $|\varrho|$ close to 1. The value of the portfolio is described by the following random variable:

$$W_p = pX_1 + (1 - p)X_2,$$

which means that among the n shares we are going to buy, there will be $n \cdot p$ shares of type A and $(1 - p)n$ shares of type B. The risk measure is the variance of the W_p

5.3 Parameters of Random Vectors

random variable, so let us calculate

$$R(p) = \text{Var} W_p = p^2\sigma_1^2 + 2p(1-p)\varrho\sigma_1\sigma_2 + (1-p)^2\sigma_2^2.$$

To minimize the risk, it is now sufficient to set $p_0 \in (0, 1)$ for which the function $R(p)$ takes the smallest possible value. Simple calculus using the derivative leads to the solution:

$$p_0 = \frac{\sigma_2^2 - \varrho\sigma_1\sigma_2}{\sigma_1^2 - 2\varrho\sigma_1\sigma_2 + \sigma_2^2}.$$

Only now can we buy the stocks. For example, if $\sigma_1 = 1$, $\sigma_2 = 2$, and $\varrho = -0.5$, we get $p_0 = 5/7$, so that for every seven shares purchased there should be five shares of type A.

5.3.3 Exercises

281. The correlation factor of X and Y is equal to α. Find the correlation factor of $Z = aX + b$ and $W = cY + d$, where $a, b, c, d \in \mathbb{R}$. What values can $\varrho(Z, W)$ take?

282. The random variables X and Y are independent, identically distributed with expected value M and variance σ^2. Find the correlation factor of the variables $Z = aX + b$ and $W = cY + d$, where $a, b, c, d \in \mathbb{R}$.

283. A random variable X satisfies the conditions: $\mathbf{P}\{X > 0\} = p > 0$, $\mathbf{P}\{X < 0\} = r > 0$, $\mathbf{E}X = a$ and $\mathbf{E}|X| = b$. Calculate $\text{Cov}(X, \text{sign}(X))$.

284. A two-dimensional random variable has the following distribution:

X/Y	−1	0	1
−1	0.125	0.5	0.125
1	0.125	0	0.125

Find the correlation factor of the variables X and Y. Are these variables independent?

285. A two-dimensional random variable is uniformly distributed over the square $\{(x, y) : |x| + |y| \leq 1\}$. Find the correlation factor $\varrho(X, Y)$. Are the variables X and Y independent?

286. A random vector (X, Y, Z) has a uniform distribution on the unit sphere $S_2 = \{(x, y, z) : x^2 + y^2 + z^2 = 1\}$. Prove that both variables X and Y have a uniform distribution on the interval $[-1, 1]$ (therefore, the joint distribution of X and Y is a copula on the square $[-1, 1]^2$). Calculate $\varrho(X, Y)$.

287. Show that if variables X and Y with zero-one distributions $\mathbf{P}\{X = 1\} = p_1 = 1 - \mathbf{P}\{X = 0\}$, $\mathbf{P}\{Y = 1\} = p_2 = 1 - \mathbf{P}\{Y = 0\}$ are uncorrelated, then they are independent. Does any two-point distribution have the same property?
Hint. Let $a = \mathbf{E}(XY) = \mathbf{P}\{X = 1, Y = 1\}$. It follows from the uncorrelation that $a = p_1 p_2$, and the two-dimensional distribution of the vector (X, Y) is uniquely determined.

288. A random vector (X, Y) has a uniform distribution on the unit sphere $S_1 = \{(x, y): x^2 + y^2 = 1\}$ in $\mathbb{R}^2$. Find the marginal distributions and the correlation matrix for this vector.

289. A random vector (X, Y) has a uniform distribution on the set $S_1 = \{(x, y): |x| + |y| = 1\}$ in $\mathbb{R}^2$. Find the marginal distributions and the correlation matrix for this vector.

290. A random vector (X, Y) has a uniform distribution on the set $S_1 = \{(x, y) \in [-1, 1]^2 : \max\{|x|, |y|\} = 1\}$. Find the marginal distributions and the correlation matrix for this vector.

291. A random vector (X, Y) has a uniform distribution on the interior of the ellipse with center at the origin and semi-axes a, b. Find the marginal distributions and the correlation matrix for this vector.

292. A random vector (X, Y) has a uniform distribution on a square with side a whose diagonals are contained in the axes of the coordinate system. Find the marginal distributions and the correlation matrix for this vector.

293. A random vector (X, Y) has a uniform distribution on $[-1, 1]^2$. Find the marginal distributions and the correlation matrix for this vector.

5.4 Multivariate Normal Distribution

We say that the random vector $\mathbf{X} = (X_1, \ldots, X_n)$ has a *multivariate normal or Gaussian distribution* if there exist a vector $\mathbf{m} = (m_1, \ldots, m_n)$ and a positive definite matrix $\Sigma = (\sigma_{ij})$ of dimension $n \times n$ such that the density function of the vector $\mathbf{X}$ at the point $\mathbf{x} = (x_1, \ldots, x_n)$ is equal to:

$$f_{N(m,\Sigma)}(\mathbf{x}) = \frac{1}{(2\pi)^{n/2}} |\Sigma|^{-1/2} \exp\left\{-\frac{1}{2}(\mathbf{x} - \mathbf{m})\Sigma^{-1}(\mathbf{x} - \mathbf{m})^T\right\},$$

where $|\Sigma|$ is the determinant of the matrix Σ, Σ^{-1} is the inverse of the matrix Σ, and $\mathbf{x}^T$ is the transposed vector $\mathbf{x}$.

Since the matrix Σ is positive definite, there exists a matrix A of dimension $n \times n$ such that $\Sigma = A^T A$, $|A| = |\Sigma|^{-1/2}$. Note that $\Sigma^{-1} = (A^T A)^{-1} = A^{-1}(A^T)^{-1}$. Now, we can find the vector of expected values and the covariance matrix for the

5.4 Multivariate Normal Distribution

random vector $\mathbf{Y} = (\mathbf{X} - \mathbf{m})A^{-1}$. Of course, $Y_i = \sum_{j=1}^{n}(X_j - m_j)c_{ij}$ if (c_{ij}) denotes the matrix A^{-1}. Hence, we obtain

$$\mathbf{E}Y_i = \int_{\mathbb{R}^n} \cdots \int \sum_{j=1}^{n}(x_j - m_j)c_{ij} \cdot f_{N(m,\Sigma)}(\mathbf{x})\, dx_1 \ldots dx_n$$

$$= \int_{\mathbb{R}^n} \cdots \int y_i \frac{|\Sigma|^{-1/2}}{(2\pi)^{n/2}} \exp\left\{-\frac{1}{2}\mathbf{y}A\Sigma^{-1}A^T\mathbf{y}^T\right\} |A|\, dy_1 \ldots dy_n$$

$$= \int_{\mathbb{R}^n} \cdots \int y_i \frac{1}{(2\pi)^{n/2}} \exp\left\{-\frac{1}{2}\mathbf{y}\mathbf{y}^T\right\} dy_1 \ldots dy_n$$

$$= \int_{\mathbb{R}} y_i \frac{1}{\sqrt{2\pi}} e^{-y_i^2/2}\, dy_i \prod_{j \neq i} \int_{\mathbb{R}} \frac{1}{\sqrt{2\pi}} e^{-y_j^2/2}\, dy_j = 0$$

because the first factor in the final product of the integrals is zero. Hence, the vector of expectations of the vector $\mathbf{Y}$ is also zero. Therefore, $\mathbf{EX} = A(\mathbf{EY}) + \mathbf{m} = \mathbf{m}$. In a similar way, we calculate the covariance for the vector X. First, we calculate $\mathrm{Cov}(Y_i, Y_j)$:

$$\mathrm{Cov}(Y_i, Y_j)$$

$$= \int_{\mathbb{R}^n} \cdots \int \left(\sum_{k=1}^{n}(x_k - m_k)c_{ik}\right)\left(\sum_{l=1}^{n}(x_l - m_l)c_{jl}\right) f_{N(m,\Sigma)}(\mathbf{x})\, dx_1 \ldots dx_n$$

$$= \int_{\mathbb{R}^n} \cdots \int y_i y_j \frac{1}{(2\pi)^{n/2}} \exp\left\{-\frac{1}{2}\sum_{k=1}^{n} y_k^2\right\} dy_1 \ldots dy_n.$$

We can now see that $\mathrm{Cov}(Y_i, Y_i) = \mathrm{Var}\, Y_i = 1$ and $\mathrm{Cov}(Y_i, Y_j) = 0$ if $i \neq j$. Let I_n denote the n-dimensional identity matrix. Since $\mathbf{X} - \mathbf{m} = A\mathbf{Y}$, $\mathrm{Cov}(X_i, X_j)$ is the expected value of the (i,j)-th element of the matrix $A\mathbf{Y}(A\mathbf{Y})^T = A\mathbf{Y}\mathbf{Y}^T A^t$. Since, as we have seen in the case of random vectors, the expected value of the matrix of random variables is equal to the matrix of expected values, we finally arrive at

$$\mathbf{E}(\mathbf{X} - \mathbf{m})(\mathbf{X} - \mathbf{m})^T = \mathbf{E}\left(A\mathbf{Y}\mathbf{Y}^T A^T\right) = A(\mathbf{E}\mathbf{Y}\mathbf{Y}^T)A^T = AI_n A^T = \Sigma.$$

It follows from the above considerations that the vector $\mathbf{X}$ with multidimensional normal distribution and density $f_{N(m,\Sigma)}$ has expected value vector $\mathbf{EX} = \mathbf{m}$ and covariance matrix Σ. We will use the notation $X \sim N(m, \Sigma)$.

Consider the two-dimensional Gaussian random vector $\mathbf{X} = (X_1, X_2)$ with expectation $m = (m_1, m_2)$ and such that $\mathrm{Var}\, X_1 = \sigma_1^2$, $\mathrm{Var}\, X_2 = \sigma_2^2$, and $\mathrm{Cov}(X_1, X_2) = \rho\sigma_1\sigma_2$, where ρ is the correlation factor of variables X_1, X_2. Then, using the explicit form of the matrix Σ^{-1}, we have

$$f(\mathbf{x}) = \frac{1}{2\pi\sigma_1\sigma_2\sqrt{1-\rho^2}} \exp\left\{-\frac{1}{2}(\mathbf{x}-\mathbf{m})\Sigma^{-1}(\mathbf{x}-\mathbf{m})^T\right\} = \frac{1}{2\pi\sigma_1\sigma_2\sqrt{1-\rho^2}}$$

$$\exp\left\{\frac{-1}{2(1-\rho^2)}\left[\frac{(x_1-m_1)^2}{\sigma_1^2} - 2\rho\frac{(x_1-m_1)(x_2-m_2)}{\sigma_1\sigma_2} + \frac{(x_2-m_2)^2}{\sigma_2^2}\right]\right\}.$$

5.4.1 Exercises

294. Prove that for a Gaussian random vector (X, Y), un-correlation is equivalent to independence.
295. A random vector $(X_1, \ldots, X_n)$ has a Gaussian distribution with expectation zero and the covariance matrix $\Sigma = (\sigma_{i,j})$. Find the distribution of the random variable $Y = \sum_{k=1}^n a_k X_k$, where $a_1, \ldots, a_k \in \mathbb{R}$.
296. The covariance matrix of a symmetric Gaussian random vector (X, Y) equals

$$\Sigma = \begin{pmatrix} 2 & 1 \\ 1 & 1 \end{pmatrix}.$$

   Calculate $\mathbf{P}\{\omega : X(\omega) > Y(\omega)\}$.
297. Let (X, Y) be a symmetric Gaussian vector with the independent identically distributed components of distribution $N(0, 1)$.

   (a) Find the distribution of $R = \sqrt{X^2 + Y^2}$.
   (b) Prove that the vector $U = \left(\frac{X}{\sqrt{X^2+Y^2}}, \frac{Y}{\sqrt{X^2+Y^2}}\right)$ has a uniform distribution on the unit sphere $S = \{(x, y) : x^2 + y^2 = 1\}$.
   (c) Show that R and U are independent.

298. Let $g(x)$ be an odd function disappearing outside the interval $[-1, 1]$ and such that $|g(x)| < (2\pi e)^{-1/2}$. Show that the function

$$f(x, y) = \frac{1}{2\pi}e^{-\frac{x^2+y^2}{2}} + g(x)g(y)$$

   is the density of a two-dimensional distribution, which is not a two-dimensional Gaussian distribution, but its marginal distributions are Gaussian.
299. A random vector (X, Y) has a two-dimensional Gaussian distribution with mean zero and covariance matrix I_2. Calculate the probability that (X, Y) falls into the area A of Lebesgue measure π in $\mathbb{R}$ if:

   (a) A is a circle centered at the origin of the coordinate system;
   (b) A is a square with its center at the origin and sides parallel to the axis;

(c) A is a rectangle with ratio of sides $10:1$, center at the origin and sides parallel to the axis.

300. For the random vector (X, Y) with density function

$$f(x, y) = C \exp\{-4x^2 - 6xy - 9y^2\},$$

find the constant C, the vector of expected values and the covariance matrix.

301. A random vector $(X_1, \ldots, X_n)$ has a multivariate Gaussian distribution with a mean of zero and covariance matrix I_n. Find the distribution of the variable

$$Y = \sqrt{X_1^2 + \cdots + X_n^2}.$$

Chapter 6
Characteristic Functions

6.1 Definition and Basic Properties

Definition 6.1 The *characteristic function* of a random variable X is defined by the formula:

$$\varphi(t) = \mathbf{E} e^{itX(\omega)}.$$

If the variable X has distribution $\mathbf{P}_X$ and distribution function F, it can also be written as

$$\varphi(t) = \int_{\mathbb{R}} e^{itx} \, \mathbf{P}_X(dx) = \int_{\mathbb{R}} e^{itx} \, dF(x).$$

If the variable X has a discrete distribution and takes values x_i, $i = 1, 2, \ldots$, with probabilities $p_i = \mathbf{P}(X = x_i)$, then its characteristic function is equal to

$$\varphi(t) = \sum_{i=1}^{\infty} e^{itx_i} p_i.$$

If the variable X is of continuous type and has density $f(x)$, then its characteristic function can be written as

$$\varphi(t) = \int_{-\infty}^{\infty} e^{itx} f(x) \, dx.$$

The latter formula means that in the case when the measure $\mathbf{P}_X$ is defined by the density f, the characteristic function is the Fourier transform of the function f. Similarly, we will sometimes say that the characteristic function is the Fourier transform of the distribution (measure) $\mathbf{P}_X$. At times, to emphasize that a function is a characteristic function of the variable X or Y, we will use the notation $\varphi_X(t)$ or $\varphi_Y(t)$.

Example 6.2 Assume that $\mathbf{P}\{X=1\} = p$, $\mathbf{P}\{X=0\} = q = 1-p$. Then,

$$\varphi(t) = \mathbf{E}e^{itX} = pe^{it} + q.$$

Example 6.3 If X is of continuous type with density $f(x) = \frac{1}{2}\exp\{-|x|\}$, then its characteristic function has the following form:

$$\varphi(t) = \mathbf{E}e^{itX} = \int_{-\infty}^{\infty} e^{itx} f(x)\,dx = \int_{0}^{\infty} \cos(tx)e^{-x}\,dx = \frac{1}{1+t^2}.$$

Theorem 6.4 *The characteristic function $\varphi(t)$ of the random variable X has the following properties:*

(1) $|\varphi(t)| \leq \varphi(0) = 1$;
(2) $\varphi(-t) = \overline{\varphi(t)} = \varphi_{-X}(t)$;
(3) $\varphi_{aX+b}(t) = e^{itb}\varphi_X(at)$;
(4) *if X, Y are independent then $\varphi_{aX+bY}(t) = \varphi_X(at)\varphi_Y(bt)$.*

Proof

(1) It should be easily seen that $\varphi(0) = \mathbf{E}e^0 = \mathbf{E}1 = 1$. Hence,

$$|\varphi(t)| = |\mathbf{E}e^{itX}| \leq \mathbf{E}|e^{itx}| = \mathbf{E}1 = \varphi(0).$$

(2) Now, all that needs to be done is to apply the formula $e^{iu} = \cos u + i \sin u$:

$$\varphi(-t) = \mathbf{E}\cos(-tX) + i\mathbf{E}\sin(-tX) = \overline{\mathbf{E}\cos(tX) + i\mathbf{E}\sin(tX)} = \overline{\varphi(t)}.$$

On the other hand, $\varphi(-t) = \mathbf{E}e^{it(-X)} = \varphi_{-X}(t)$.
(3) $\varphi_{aX+b}(t) = \mathbf{E}e^{itaX+itb} = e^{itb}\mathbf{E}e^{itaX} = e^{itb}\varphi_X(at)$.
(4) If the variables X, Y are independent, then also e^{itX} and e^{itY} are independent. Thus,

$$\mathbf{E}\exp\{it(aX+bY)\} = \mathbf{E}e^{itaX} \cdot \mathbf{E}e^{itbY} = \varphi_X(at)\varphi_Y(bt).$$

$\square$

Example 6.5 Consider the random variable X of the Bernoulli distribution with parameters n and p. Its characteristic function can be calculated directly from the definition:

$$\varphi(t) = \sum_{k=0}^{n} e^{itk}\mathbf{P}\{X=k\} = \sum_{k=0}^{n} e^{itk}\binom{n}{k}p^k q^{n-k}.$$

6.1 Definition and Basic Properties

However, it is easier to note that the random variable X has the same distribution as $\sum_{k=1}^{n} Y_k$, for independent random variables $Y_1, \ldots, Y_n$, with the same zero-one distribution $\mathbf{P}\{Y_k = 1\} = p, \mathbf{P}\{Y_k = 0\} = q$. By property 4) and mathematical induction, we obtain

$$\varphi(t) = \prod_{k=1}^{n} \varphi_{Y_k}(t) = \left(pe^{it} + q\right)^n.$$

Theorem 6.6 *Each characteristic function is uniformly continuous on the real line.*

Proof First, note that

$$|\varphi(t+h) - \varphi(t)| = |\mathbf{E}e^{itX}(e^{ihX} - 1)| \leqslant \mathbf{E}|e^{ihX} - 1|.$$

The function $g_h(x) = |e^{ihx} - 1|$ is bounded by 2 along the real line. Furthermore, for any $m > 0$, the values of $g_h(x)$ converge to zero uniformly on $[-m, m]$ if h converges to zero. For $\varepsilon > 0$, we choose m large enough that $\mathbf{P}\{|X(\omega)| > m\} < \varepsilon/4$. Next, let h_0 be small enough so that for each $h \in (0, h_0)$ and each $x \in [-m, m]$ the condition $g_h(x) < \varepsilon/2$ is satisfied. Then, for $h \in (0, h_0)$, we get

$$|\varphi(t+h) - \varphi(t)| \leqslant \int_{-m}^{m} |e^{ihx} - 1| \, dF(x) + \int_{|x|>m} |e^{ihX} - 1| \, dF(x)$$

$$\leqslant \varepsilon/2 \mathbf{P}(|X(\omega)| \leqslant m) + 2\mathbf{P}(|X(\omega)| > m) < \varepsilon.$$

□

6.1.1 Exercises

302. Prove that if the characteristic function $\varphi(t)$ of the distribution function F is even, then

$$\varphi(t) = \int_{-\infty}^{\infty} \cos tx \, dF(x).$$

303. Prove that the following cannot be characteristic functions of any probability distribution:

    (a) $\varphi(t) = e^{i|t|}$;
    (b) $\varphi(t) = a \cos t + b \sin t$ for $a, b \in \mathbb{R} \setminus \{0\}$.

304. Let $\Omega = [0, 1]$ with the σ-field of Borel sets and the Lebesgue probability measure. Find the characteristic functions of the following random variables:

$$X(\omega) = 2\omega - \mathbf{1}_{(\frac{1}{2}, 1]}; \quad Y(\omega) = 1 - \mathbf{1}_{(\frac{1}{3}, \frac{2}{3}]}; \quad Z(\omega) = \ln \omega \mathbf{1}_{(0, 1]}.$$

305. Calculate the characteristic functions of the absolutely continuous random variables with the following densities:

(a) $f(x) = 2x\mathbf{1}_{[0,1]}$; (b) $f(x) = 4x\mathbf{1}_{[0,\frac{1}{2}]} + (4 - 4x)\mathbf{1}_{[\frac{1}{2},1]}$;

(c) $f(x) = \dfrac{3}{2}x^2\mathbf{1}_{[-1,1]}$.

306. X and Y are independent, asymmetric random variables. Could the variable $Z = X + Y$ be symmetric?

307. Find the characteristic function of the distribution $N(m, \sigma)$. Prove that if X and Y are independent random variables with normal distributions (not necessarily identical), then for any $a, b \in \mathbb{R}$, the variable $Z = aX + bY$ has a normal distribution as well.

308. Calculate the characteristic functions of the following distributions: (a) uniform on the interval $[a, b]$; (b) Bernoulli with parameters n, p; (c) Poisson with parameter λ; (d) exponential with parameter a; (e) discrete geometric; (f) waiting time for the k-th success in n Bernoulli trials.

309. Independent random variables $X_1, \ldots, X_n$ have the same Cauchy distribution with parameters $a > 0$ and $m \in \mathbb{R}$. Show that the variable $Y = \sum_{k=1}^{n} X_k$ has a Cauchy distribution, too.

310. Independent random variables $X_1, \ldots, X_n$ have gamma distributions: $\Gamma(p_1, a), \ldots, \Gamma(p_n, a)$. Find the distribution of $Y = \sum_{k=1}^{n} X_k$.

311. Let $\varphi_1(t), \ldots, \varphi_n(t)$ be the characteristic functions of independent random variables $X_1, \ldots, X_n$ and let the numbers $\alpha_1, \ldots, \alpha_n$ be positive and such that $\alpha_1 + \cdots + \alpha_n = 1$. Prove that $\varphi(t) = \alpha_1\varphi_1(t) + \cdots + \alpha_n\varphi_n(t)$ is a characteristic function as well.

312. Let $\varphi(t)$ be the characteristic function of a distribution function F. Prove that the following functions are characteristic functions as well:

$$\varphi_1(t) = \frac{2}{2 - \varphi(t)} - 1; \qquad \varphi_2(t) = \Re(\varphi(t)); \qquad \varphi_3(t) = |\varphi(t)|^2.$$

313. Prove that if φ is a characteristic function, then the function ϕ defined by the formula

$$\phi(t) = \frac{1}{t}\int_0^t \varphi(x)\,dx$$

is also a characteristic function.

314. The variable X has a standard normal distribution. Find the characteristic function of the variable $Y = X^2$.

315. Let the one-to-one function $F: \mathbb{R} \to [0, 1]$ be the distribution function of a random variable X. Find the characteristic function of $Y = \ln F(X)$.

6.2 Relations Between Distribution, Characteristic Function and Moments of Random Variables

Theorem 6.7 *Let φ be the characteristic function of a random variable with distribution function F. Then:*

(a) *for every $a < b$*

$$G = \lim_{R \to \infty} \frac{1}{2\pi} \int_{-R}^{R} \frac{e^{-ita} - e^{-itb}}{it} \varphi(t) \, dt = \mathbf{P}((a,b)) + \frac{1}{2}\mathbf{P}(\{a\}) + \frac{1}{2}\mathbf{P}(\{b\});$$

(b) *if a and b are continuity points for the distribution function F, then*

$$\lim_{R \to \infty} \frac{1}{2\pi} \int_{-R}^{R} \frac{e^{-ita} - e^{-itb}}{it} \varphi(t) \, dt = F(b) - F(a);$$

(c) *if $\int_{-\infty}^{\infty} |\varphi(t)| \, dt < \infty$, then F is of continuous type with density $f(\cdot)$, where*

$$f(x) = \frac{1}{2\pi} \int_{-\infty}^{\infty} e^{-itx} \varphi(t) \, dt.$$

Proof Consider the integral

$$G(R) = \frac{1}{2\pi} \int_{-R}^{R} \frac{e^{-ita} - e^{-itb}}{it} \varphi(t) \, dt$$

$$= \frac{1}{2\pi} \int_{-R}^{R} \frac{e^{-ita} - e^{-itb}}{it} \int_{-\infty}^{\infty} e^{itx} \, dF(x) \, dt.$$

The function under the integral is bounded here because

$$\left| \frac{e^{-ita} - e^{-itb}}{it} e^{itx} \right| = \left| \frac{e^{-ita} - e^{-itb}}{it} \right| = \left| \int_{a}^{b} e^{itx} \, dx \right| \leq \int_{a}^{b} dx = b - a.$$

By Fubini's Theorem, we can change the order of integration and get

$$G(R) = \frac{1}{2\pi} \int_{-\infty}^{\infty} \int_{-R}^{R} \frac{e^{it(x-a)} - e^{it(x-b)}}{it} \, dt \, dF(x).$$

We have $e^{-itc} = \cos(tc) + i\sin(tc)$. Noting that the integral of the function $\cos(tc)/it$ on the set $[-R, R]$ disappears (as does any integral of an odd function over a symmetric set), we obtain:

$$G(R) = \frac{1}{2\pi} \int_{-\infty}^{\infty} \left\{ 2\int_0^R \frac{\sin t(x-a)}{t} dt - 2\int_0^R \frac{\sin t(x-b)}{t} dt \right\} dF(x)$$

$$= \frac{1}{\pi} \int_{-\infty}^{\infty} \left\{ \int_0^{R(x-a)} \frac{\sin y}{y} dy - \int_0^{R(x-b)} \frac{\sin y}{y} dy \right\} dF(x)$$

$$= \frac{1}{\pi} \int_{-\infty}^{\infty} \int_{R(x-b)}^{R(x-a)} \frac{\sin y}{y} dy\, dF(x).$$

We will now prove that the function $\int_0^z \frac{\sin y}{y} dy$ is bounded. Due to the fact that the integrand function is even, we only need to consider $z > 0$. If $0 < z < \frac{\pi}{2}$, then

$$\left| \int_0^z \frac{\sin y}{y} dy \right| \leq \int_0^z \left|\frac{\sin y}{y}\right| dy \leq \int_0^z dy = z < \frac{\pi}{2}.$$

For $z > \frac{\pi}{2}$, integrating by parts, we obtain

$$\left| \int_0^z \frac{\sin y}{y} dy \right| \leq \left| \int_0^{\pi/2} \frac{\sin y}{y} dy \right| + \left|\frac{\cos y}{y}\right|_{\pi/2}^z + \left| \int_{\pi/2}^z \frac{\cos y}{y^2} dy \right|$$

$$\leq \frac{\pi}{2} + \frac{1}{z} + \frac{2}{\pi} + \int_{\pi/2}^\infty \frac{1}{y^2} dy$$

$$\leq \frac{\pi}{2} + \frac{6}{\pi} = \text{const.}$$

From the Lebesgue Dominated Convergence Theorem, we get:

$$\lim_{R \to \infty} G(R) = \lim_{R \to \infty} \frac{1}{\pi} \int_{-\infty}^{\infty} \int_{R(x-b)}^{R(x-a)} \frac{\sin y}{y} dy\, dF(x)$$

$$= \frac{1}{\pi} \int_{-\infty}^{\infty} \left\{ \lim_{R \to \infty} \int_{R(x-b)}^{R(x-a)} \frac{\sin y}{y} dy \right\} dF(x).$$

Now, we need to consider the following cases:

- if $a < x < b$, then

$$\lim_{R \to \infty} \frac{1}{\pi} \int_{R(x-b)}^{R(x-a)} \frac{\sin y}{y} dy = \frac{1}{\pi} \int_{-\infty}^{\infty} \frac{\sin y}{y} dy = 1;$$

6.2 Relations Between Distribution, Characteristic Function and Moments

- if $x > b$, then $\lim_{R \to \infty} R(x - b) = \infty$, which due to the integrability of $\frac{\sin y}{y}$, leads to

$$\lim_{R \to \infty} \frac{1}{\pi} \int_{R(x-b)}^{R(x-a)} \frac{\sin y}{y} \, dy = 0;$$

- if $x < a$, as before $\lim_{R \to \infty} R(x - a) = -\infty$, and

$$\lim_{R \to \infty} \frac{1}{\pi} \int_{R(x-b)}^{R(x-a)} \frac{\sin y}{y} \, dy = 0;$$

- for $x = a$, we get:

$$\lim_{R \to \infty} \frac{1}{\pi} \int_{-R(b-a)}^{0} \frac{\sin y}{y} \, dy = \frac{1}{\pi} \int_{-\infty}^{0} \frac{\sin y}{y} \, dy = \frac{1}{2};$$

- if $x = b$, then

$$\lim_{R \to \infty} \frac{1}{\pi} \int_{0}^{R(b-a)} \frac{\sin y}{y} \, dy = \frac{1}{\pi} \int_{0}^{\infty} \frac{\sin y}{y} \, dy = \frac{1}{2}.$$

Hence, it is easy to deduce that

$$\lim_{R \to \infty} G(R) = \int_{-\infty}^{\infty} \left\{ \mathbf{1}_{(a,b)}(x) + \frac{1}{2} \mathbf{1}_{\{a\}}(x) + \frac{1}{2} \mathbf{1}_{\{b\}}(x) \right\} dF(x)$$

$$= \mathbf{P}((a,b)) + \frac{1}{2} \mathbf{P}(\{a\}) + \frac{1}{2} \mathbf{P}(\{b\}).$$

The property (b) is an obvious consequence of (a) because, by the continuity of the distribution function at a and b, we have $\mathbf{P}(\{a\}) = \mathbf{P}(\{b\}) = 0$. To prove (c), let us assume that $\int_{-\infty}^{\infty} |\varphi(t)| \, dt < \infty$. Then, the following function

$$f(x) = \frac{1}{2\pi} \int_{-\infty}^{\infty} e^{-itx} \varphi(t) \, dt$$

is continuous and integrable on every interval $[a, b]$. Hence,

$$\int_{a}^{b} f(x) \, dx = \int_{a}^{b} \frac{1}{2\pi} \int_{-\infty}^{\infty} e^{-itx} \varphi(t) \, dt \, dx$$

$$= \frac{1}{2\pi} \int_{-\infty}^{\infty} \varphi(t) \int_{a}^{b} e^{-itx} \, dx \, dt$$

$$= \frac{1}{2\pi} \int_{-\infty}^{\infty} \varphi(t) \frac{e^{-ita} - e^{-itb}}{it} \, dt$$

$$= \lim_{R \to \infty} \frac{1}{2\pi} \int_{-R}^{R} \varphi(t) \frac{e^{-ita} - e^{-itb}}{it} \, dt$$

$$= \mathbf{P}((a,b)) + \frac{1}{2}\mathbf{P}(\{a\}) + \frac{1}{2}\mathbf{P}(\{b\}).$$

Since f was a continuous function, the obtained integral has to be a continuous function of a and b, and

$$\forall a < b \qquad \int_a^b f(x) \, dx = F(b) - F(a),$$

which is equivalent to f being the density of a random variable with distribution function F. □

As a simple consequence of the above theorem, we get a very important uniqueness theorem:

Theorem 6.8 *The characteristic function uniquely determines the distribution of the random variable.*

Example 6.9 The function $\varphi(t) = e^{-|t|}$ is an integrable characteristic function. According to Theorem 6.7, the corresponding probability distribution has the following density function:

$$f(x) = \frac{1}{2\pi} \int_{\mathbb{R}} e^{-itx} e^{-|t|} \, dt = \frac{1}{2\pi} \int_{\mathbb{R}} \cos(tx) e^{-|t|} \, dt = \frac{1}{\pi(1+x^2)}.$$

The uniqueness of the characteristic function shows that $\varphi(t)$ is a characteristic function of the Cauchy distribution with parameters $a = 1$ and $m = 0$.

Theorem 6.10 *The characteristic function $\varphi_X(t)$ is real if and only if the corresponding random variable X is symmetric, that is, if X and $-X$ have the same distribution.*

Proof If both X and $-X$ have the same distribution, then, of course, $\varphi_X(t) = \varphi_{-X}(t)$. Equivalently, $\varphi_{-X}(t) = \varphi_X(-t) = \overline{\varphi(t)}$. Hence, it is easy to get that $\Im\mathfrak{m}(\varphi_X(t)) \equiv 0$.

Suppose that $\varphi_X(t)$ is a real function. Then, we have $\varphi_{-X}(t) = \varphi_X(-t) = \overline{\varphi_X(t)} = \varphi_X(t)$. Since the characteristic function determines the distribution uniquely, we conclude that the random variables X and $-X$ have identical distributions. □

6.2 Relations Between Distribution, Characteristic Function and Moments

Theorem 6.11 *If for some $n \geq 1$, $\mathbf{E}|X|^n < \infty$, then for every $r \leq n$, the derivative $\varphi^{(r)}(t)$ exists, and*

$$\varphi^{(r)}(t) = \int_{-\infty}^{\infty} (ix)^r e^{itx} \, dF(x);$$

$$\mathbf{E}X^r = \frac{\varphi^{(r)}(0)}{i^r}.$$

Proof If $\mathbf{E}|X|^n < \infty$, then, of course, $\mathbf{E}|X|^r < \infty$ for every $0 < r \leq n$. Consider the difference quotient

$$\frac{\varphi(t+h) - \varphi(t)}{h} = \mathbf{E}\left[e^{itX} \frac{e^{ihX} - 1}{h}\right].$$

Since $|e^{ihX} - 1| \leq |hX|$ and $\mathbf{E}|X| < \infty$, there exists the limit of the difference quotient when h tends to zero, and

$$\varphi'(t) = \lim_{h \to 0} \mathbf{E}\left[e^{itX} \frac{e^{ihX} - 1}{h}\right] = \mathbf{E}\left[e^{itX} \lim_{h \to 0} \frac{e^{ihX} - 1}{h}\right] = \mathbf{E}\left[iXe^{itX}\right].$$

By mathematical induction, we get

$$\varphi^{(r)}(t) = \int_{-\infty}^{\infty} (ix)^r e^{itx} \, dF(x),$$

and applying this formula to the value of $t = 0$ completes the proof. □

Remark 6.12 The converse implication in the above theorem does not hold. It is possible for the characteristic function f of the random variable X to have a k-th derivative, but the variable X not to have a k-th moment. However, it can be proved that if $\varphi^{(2k)}$ exists, then $\mathbf{E}X^{2k} < \infty$. We use the notation $\varphi^{(k)}$ for the k-th derivative of the function φ. Here, we will only prove the following theorem:

Theorem 6.13 *Let φ be the characteristic function of the random variable X. If φ is twice differentiable at zero, then $\mathbf{E}X^2 < \infty$.*

Proof Since $|\varphi(t)| \leq \varphi(0) = 1$ for each $t \in \mathbb{R}$, then it is not possible for φ to be convex around zero. Consequently, it cannot happen that $\varphi''(0) > 0$. Applying de l'Hospital's rule, we obtain:

$$\varphi''(0) = \lim_{h \to 0} \frac{1}{2}\left[\frac{\varphi'(2h) - \varphi'(0)}{2h} + \frac{\varphi'(0) - \varphi(-2h)}{2h}\right]$$

$$= \lim_{h \to 0} \frac{\varphi'(2h) - \varphi'(-2h)}{4h} = \lim_{h \to 0} \frac{\varphi(2h) - 2\varphi(0) + \varphi(-2h)}{4h^2}$$

$$= \lim_{h \to 0} \int_{\mathbb{R}} \left(\frac{e^{itx} - e^{-itx}}{2h} \right)^2 dF(x) = -\lim_{h \to 0} \int_{\mathbb{R}} \left(\frac{\sin hx}{hx} \right)^2 x^2 \, dF(x).$$

We conclude from this that $\varphi''(0)$ is negative real. By virtue of Fatou's Lemma, we finally get

$$-\varphi''(0) = \lim_{h \to 0} \int_{\mathbb{R}} \left(\frac{\sin hx}{hx} \right)^2 x^2 \, dF(x)$$

$$\geq \int_{\mathbb{R}} \lim_{h \to 0} \left(\frac{\sin hx}{hx} x^2 \right)^2 dF(x) = \int_{\mathbb{R}} x^2 \, dF(x),$$

which was to be shown. □

6.2.1 Exercises

316. It is known that the random variable X has an atom at some point $a \in \mathbb{R}$ and $\mathbf{P}\{X = a\} > \frac{1}{2}$. Prove that the characteristic function of this variable cannot take negative values.
317. Prove that there exist normally distributed random variables X and Y whose joint distribution is not a two-dimensional normal distribution.
    Hint: Construct a random vector (X, Y) living on diagonals of the plane such that: $\mathbf{P}\{X = Y\} = \mathbf{P}\{X = -Y\} = \frac{1}{2}$.
318. Let φ be the characteristic function of a variable X. Describe the consequences of the condition $\varphi''(0) = 0$.
319. Is the function $\varphi(t) = \cos(t^2)$ a characteristic function of some probability distribution?
320. Let $X_1, X_2, \ldots$ be a sequence of independent random variables with the same uniform distribution on the interval $[-1, 1]$, and let $S_n = X_1 + \cdots + X_n$.

    (a) Find the joint density function of the vector (S_2, S_3).
    (b) Find the limit of the sequence of characteristic functions for random variables $n^{-1/2} S_n$ when $n \to \infty$.

321. Let $\mathbf{X}_1, \mathbf{X}_2, \ldots$ be independent, identically distributed, two-dimensional random vectors which take values in the integer lattice $Z^2 = \{(m, n) : m, n \in Z\}$. The variable $\mathbf{S}_n = \mathbf{X}_1 + \cdots + \mathbf{X}_n$ specifies the position of a particle after n steps, assuming that initially it was at the origin $(0, 0)$ of the coordinate system. Let us treat $\mathbf{X}_j$ as the j-th step of the particle here. Find the limit of the two-dimensional characteristic function of the variable $n^{-1/2} \mathbf{S}_n$ at the point (r, s) as $n \to \infty$ if $\varphi_{(X,Y)}(r, s) = \mathbf{E} \exp\{i(rX + sY)\}$, and

    (a) $\mathbf{X}_j$ takes four values $(0, 1), (1, 0), (-1, 0), (0, -1)$, with probability $\frac{1}{4}$ each;

(b) $\mathbf{P}\{\mathbf{X}_i = (m, n)\} = \frac{1}{9}$ for $m, n \in \{-1, 0, 1\}$;
(c) $\mathbf{P}\{\mathbf{X}_i = (-1, 0)\} = \mathbf{P}\{\mathbf{X}_i = (1, 0)\} = \frac{1}{3}$,
$\mathbf{P}\{\mathbf{X}_i = (-1, 1)\} = \mathbf{P}\{\mathbf{X}_i = (1, -1)\} = \frac{1}{6}$.

6.3 Weak Convergence of Distributions

Definition 6.14 Let $\mu, \mu_1, \mu_2, \ldots$ be a sequence of probability distributions on $(\mathbb{R}, \mathcal{B})$. We say that the sequence (μ_n) is *weakly convergent* to the distribution μ (notation $\mu_n \xrightarrow{\omega} \mu$ or $\mu_n \Rightarrow \mu$) if for every continuous and bounded function f on $\mathbb{R}$, we have

$$\lim_{n \to \infty} \int_{\mathbb{R}} f(x) \mu_n(dx) = \int_{\mathbb{R}} f(x) \mu(dx).$$

Here, the term *weak convergence* refers to the general concept from functional analysis. Each bounded and continuous function f on the line defines a linear functional over the space of measures on $\mathbb{R}$ via the formula $\mu \to \int f d\mu$. We expect that for any such functional, its values on the sequence μ_n, $n \in \mathbb{N}$, converge to the value on μ if and only if $\mu_n \Rightarrow \mu$.

In the language of the corresponding random variables, weak convergence of the distributions is called convergence *in distribution*.

Definition 6.15 Let $X, X_1, X_2, \ldots$ be a sequence of random variables with distributions $\mu, \mu_1, \mu_2, \ldots$ We say that the sequence (X_n) *converges in distribution* to the random variable X (notation $X_n \xrightarrow{d} X$) if $\mu_n \Rightarrow \mu$.

Convergence in distribution translates into a very interesting property of the distribution functions of the considered random variables.

Theorem 6.16 *Let* $X, X_1, X_2, \ldots$ *be a sequence of random variables, and let* $F, F_1, F_2, \ldots$ *be the sequence of the corresponding cumulative distribution functions. Then,* $X_n \xrightarrow{d} X$ *if and only if*

$$\lim_{n \to \infty} F_n(x) = F(x)$$

at each continuity point x of the limit distribution function F.

Proof Assume that $X_n \xrightarrow{d} X$, and let x be a continuity point of the distribution function F. For every $\varepsilon > 0$, there exists a $\delta > 0$ for which $|F(x \pm \delta) - F(x)| < \frac{\varepsilon}{2}$. Consider the following functions:

$$f_+(y) = \begin{cases} 1, & y \leq x; \\ 1 - \frac{y-x}{\delta}, & x < y \leq x+\delta; \\ 0, & y > x+\delta; \end{cases} \qquad f_-(y) = \begin{cases} 1, & y \leq x-\delta; \\ -\frac{y-x}{\delta}, & x-\delta < y \leq x; \\ 0, & y > x. \end{cases}$$

Both functions are continuous and bounded on $\mathbb{R}$. Therefore, the condition $X_n \xrightarrow{d} X$ implies the existence of an $n_0 \in \mathbb{N}$ such that for every $n \geq n_0$

$$\int f_+(y) \, dF_n(y) \leq \int f_+(y) \, dF(y) + \frac{\varepsilon}{2} \quad \text{and} \quad \int f_-(y) \, dF_n(y) \geq \int f_-(y) \, dF(y) - \frac{\varepsilon}{2}.$$

Hence, for $n \geq n_0$, we get:

$$F_n(x) = \int \mathbf{1}_{(-\infty,x)} \, dF_n(y) \leq \int f_+(y) \, dF_n(y) \leq \int f_+(y) \, dF(y) + \frac{\varepsilon}{2}$$

$$\leq \int \mathbf{1}_{(-\infty,x+\delta)} \, dF(y) + \frac{\varepsilon}{2} = F(x+\delta) + \frac{\varepsilon}{2}.$$

On the other hand, we have

$$F_n(x) = \int \mathbf{1}_{(-\infty,x)} \, dF_n(y) \geq \int f_-(y) \, dF_n(y) \geq \int f_-(y) \, dF(y) - \frac{\varepsilon}{2}$$

$$\geq \int \mathbf{1}_{(-\infty,x-\delta)} \, dF(y) - \frac{\varepsilon}{2} = F(x-\delta) - \frac{\varepsilon}{2}.$$

This leads us to:

$$F(x-\delta) - F(x) - \frac{\varepsilon}{2} \leq F_n(x) - F(x) \leq F(x+\delta) - F(x) + \frac{\varepsilon}{2},$$

and hence, it is easy to see that for $n \geq n_0$, we have $|F_n(x) - F(x)| < \varepsilon$.

To prove the converse implication, suppose that $\lim F_n(x) = F(x)$ at each continuity point of the distribution function F, and let f be a continuous function such that $\sup\{|f(x)| : x \in \mathbb{R}\} = M < \infty$. It is known that the set of discontinuity points of the cumulative distribution function is at most countable, hence in every neighborhood of each discontinuity point, some continuity points can be found.

For a fixed $\varepsilon > 0$, let us first select two numbers $a, b, a < b$, which are continuity points of the cumulative distribution function F and such that $F(a) < \frac{\varepsilon}{16M}$ and $1 - F(b) < \frac{\varepsilon}{16M}$. From the assumption of the convergence of the distribution functions (F_n) at points a and b, it follows that there is an $n_0 \in \mathbb{N}$ such that for each $n \geq n_0$,

6.3 Weak Convergence of Distributions

the following conditions hold: $F_n(a) < \frac{\varepsilon}{8M}$ and $1 - F_n(b) < \frac{\varepsilon}{8M}$. Then, for $n \geq n_0$, we get:

$$I_n \stackrel{\text{def}}{=} \left| \int f \, dF_n - \int f \, dF \right| \leq \left| \int_{[a,b]} f \, dF_n - \int_{[a,b]} f \, dF \right|$$

$$+ \int_{(-\infty,a)} |f| \, dF_n + \int_{(-\infty,a)} |f| \, dF + \int_{(b,\infty)} |f| \, dF_n + \int_{(b,\infty)} |f| \, dF$$

$$\leq \left| \int_{[a,b]} f \, dF_n - \int_{[a,b]} f \, dF \right| + \frac{3\varepsilon}{8}.$$

Since the continuous function f on the compact interval $[a, b]$ is uniformly continuous, there exists a $\delta > 0$ such that if $x, y \in [a, b]$ and $|x - y| < \delta$, then $|f(x) - f(y)| < \frac{\varepsilon}{4}$. We can divide the interval $[a, b]$ into k subintervals of length δ with division points $x_0 = a < x_1 < \cdots < x_k = b$. Points $x_1, \ldots, x_k$ are chosen from among the continuity points of the distribution function F.

Now, let us define the function f_δ on the interval $[a, b]$:

$$f_\delta(x) = \sum_{i=0}^{k-1} f(x_i) \mathbf{1}_{[x_i, x_{i+1})}(x).$$

Of course, $|f(x) - f_\delta(x)| < \frac{\varepsilon}{4}$ for every $x \in [a, b)$. Moreover,

$$I_n \leq \frac{3\varepsilon}{8} + \left| \int_{[a,b]} f \, dF_n - \int_{[a,b]} f_\delta \, dF_n \right| + \left| \int_{[a,b]} f \, dF - \int_{[a,b]} f_\delta \, dF \right|$$

$$+ \left| \int_{[a,b]} f_\delta \, dF_n - \int_{[a,b]} f_\delta \, dF \right|$$

$$\leq \frac{3\varepsilon}{8} + \int_{[a,b]} |f - f_\delta| \, dF_n + \int_{[a,b]} |f - f_\delta| \, dF + \left| \int_{[a,b]} f_\delta \, dF_n - \int_{[a,b]} f_\delta \, dF \right|$$

$$\leq \frac{3\varepsilon}{8} + \frac{2\varepsilon}{4} + \left| \int_{[a,b]} f_\delta \, dF_n - \int_{[a,b]} f_\delta \, dF \right| = \frac{7\varepsilon}{8} + \left| \int_{[a,b]} f_\delta \, dF_n - \int_{[a,b]} f_\delta \, dF \right|.$$

We still need to estimate the last difference. Note that f_δ is a simple function, thus, by the definition of the integral with respect to increments of the distribution function, we get

$$I_n \leq \frac{7\varepsilon}{8} + \left| \sum_{i=0}^{k-1} f(x_i)[F_n(x_{i+1}) - F_n(x_i)] - \sum_{i=0}^{k-1} f(x_i)[F(x_{i+1}) - F(x_i)] \right|$$

$$\leq \frac{7\varepsilon}{8} + \sum_{i=0}^{k-1} |f(x_i)| \big[|F_n(x_{i+1}) - F(x_{i+1})| + |F_n(x_i) - F(x_i)| \big].$$

From the fact that $x_0, x_1, \ldots, x_k$ are the continuity points of the cumulative distribution function F and the assumptions on the convergence of the sequence (F_n), it follows that there exists an $n_1 \in \mathbb{N}$ such that for every $n \geq n_1$

$$|F_n(x_i) - F(x_i)| < \frac{\varepsilon}{16kM}, \quad i = 0, 1, \ldots, k.$$

Hence, it follows that for every $n \geq \max\{n_0, x_1\}$,

$$I_n \leq \frac{7\varepsilon}{8} + k \cdot M \cdot \frac{\varepsilon}{16kM} = \varepsilon,$$

which was to be shown. $\square$

The Lévy–Cramér Continuity Theorem plays a key role in problems of convergence with respect to distribution. In order to prove it, we will use the following lemma:

Lemma 6.17 *Let F be a distribution function on $\mathbb{R}$ with corresponding characteristic function φ. Then, for every $u > 0$,*

$$F(2/u) - F(-2/u) \geq 1 - \frac{1}{u} \int_{-u}^{u} (1 - \varphi(s)) \, ds.$$

Proof Note that

$$\frac{1}{u} \int_{-u}^{u} (1 - \varphi(s)) \, ds = \frac{1}{u} \int_{-\infty}^{\infty} \int_{-u}^{u} \left(1 - e^{isx}\right) ds \, dF(x)$$

$$= 2 \int_{-\infty}^{\infty} \left(1 - \frac{\sin ux}{ux}\right) dF(x)$$

$$\geq 2 \int_{(-\infty, -2/u)} \frac{1}{2} dF(x) + 2 \int_{[2/u, \infty)} \frac{1}{2} dF(x)$$

$$= F(-2/u) + 1 - F(2/u).$$

The result of the lemma follows easily from these calculations. $\square$

Theorem 6.18 (Lévy–Cramér Continuity Theorem) *Let (X_n) be a sequence of random variables and (φ_n) the corresponding sequence of their characteristic functions. If $\varphi_n(t) \to \varphi(t)$ for every $t \in \mathbb{R}$ and the function φ is continuous at zero, then φ is a characteristic function of some random variable X, and $X_n \xrightarrow{d} X$.*

Remark 6.19 Of course, the converse implication is also true.

If $\varphi, \varphi_1, \ldots$ are characteristic functions of random variables $X, X_1, X_2, \ldots$ and $X_n \xrightarrow{d} X$, then $\varphi_n(t) \to \varphi(t)$ for every $t \in \mathbb{R}$.

6.3 Weak Convergence of Distributions

To see this, note that

$$\varphi_n(t) = \int_\Omega e^{itX_n}\,d\mathbf{P} = \int_\mathbb{R} \cos(tx)\,dF_n(x) + i\int_\mathbb{R} \sin(tx)\,dF_n(x),$$

where F_n is the distribution function of the variable X_n. Both functions $\cos(tx)$ and $\sin(tx)$ are bounded and continuous on $\mathbb{R}$, so the convergence of $\varphi_n(t)$ to $\varphi(t)$ follows directly from the definition of convergence in distribution.

Proof of the Lévy–Cramér Theorem Let $F_1, F_2 \ldots$ be distribution functions of the variables $X_1, X_2, \ldots$. The proof will be carried out in a few steps.

Step 1 Let us set $\varepsilon > 0$. Since the function φ is continuous at zero and $\varphi(0) = 1$, the values of $\varphi(t)$ in a sufficiently small neighborhood of zero are only slightly different from 1, and we can choose $u > 0$ such that

$$\frac{1}{u}\int_{-u}^{u}(1-\varphi(s))\,ds \leq \frac{\varepsilon}{2}.$$

From the assumption of the convergence of the sequence $\varphi_n(t)$ for each $t \in \mathbb{R}$ and the Lebesgue Dominated Convergence Theorem, it follows that there exists an $n_0 \in \mathbb{N}$ such that for each $n \geq n_0$

$$\frac{1}{u}\int_{-u}^{u}(1-\varphi_n(s))\,ds \leq \varepsilon.$$

By Lemma 6.17, we get that for $n \geq n_0$

$$F_n(2/u) - F_n(-2/u) \geq 1 - \varepsilon.$$

Step 2 Now let $S_u = \{s_n : n \in \mathbb{N}\}$ be a sequence comprising all rational numbers contained in the interval $I_u = [-2/u, 2/u]$. Using the diagonal method, from the sequence $(F_n)_{n\in\mathbb{N}}$ we can choose a subsequence $(F_{n,n})_{n\in\mathbb{N}}$ which converges to a certain value denoted by $F_u(s_i)$ at every point $s_i \in S_u$. For this, let us first consider the sequence of numbers $F_n(s_1), n \in \mathbb{N}$. This sequence is bounded on both sides, so it contains a subsequence converging to the point which we will denote by $F_u(s_1) \in [0, 1]$. Hence, there exists a subsequence $(F_{1,n})_{n\in\mathbb{N}}$ of the sequence $(F_n)_{n\in\mathbb{N}}$ such that

$$\lim_{n\to\infty} F_{1,n}(s_1) = F_u(s_1).$$

Considering now the sequence $F_{1,n}(s_2), n \in \mathbb{N}$, we can find a subsequence $(F_{2,n})_{n\in\mathbb{N}}$ of the sequence $(F_{1,n})_{n\in\mathbb{N}}$ and a number $F_u(s_2) \in [0, 1]$ such that

$$\lim_{n\to\infty} F_{2,n}(s_2) = F_u(s_2).$$

Continuing this construction until the set S_u is empty, we get the table of distribution functions $(F_{k,n})$, $n, k \in \mathbb{N}$ and a sequence of numbers $(F_u(s_k))_{k \in \mathbb{N}}$ such that $(F_{k,n})_{n \in \mathbb{N}}$ is a subsequence of the sequence $(F_{m,n})_{n \in \mathbb{N}}$ if only $m \leqslant k$ and

$$\lim_{n \to \infty} F_{k,n}(s_m) = F_u(s_m) \quad \text{for } m \leqslant k, \ m, k \in \mathbb{N}.$$

Hence, it follows that the sequence $(F_{n,n})_{n \in \mathbb{N}}$ built of elements lying on the diagonal of this table satisfies the condition

$$\lim_{n \to \infty} F_{n,n}(s_k) = F_u(s_k) \quad \text{for every } s_k \in S_u.$$

Step 3 We can now define a new distribution function F_u, where

$$F_u(t) = \begin{cases} 0, & t \leqslant -2/u; \\ \sup\{F(s_k) : s_k < t, s_k \in S_u\}, & -2/u < t \leqslant 2/u; \\ 1, & t > 2/u. \end{cases}$$

Obviously, F_u is left-continuous and nondecreasing. Moreover, for each point $t \in (-2/u, 2/u) = I_u$ which is a continuity point of F_u, we have

$$\lim_{n \to \infty} F_{n,n}(t) = F(t).$$

Indeed, if $t \in I_u$ is a continuity point of F_u and $r_1, r_2 \in S_u$ are such that $r_1 < t < r_2$, then $F_{n,n}(r_1) \leqslant F_{n,n}(t) \leqslant F_{n,n}(r_2)$. Hence,

$$\liminf_{n \to \infty} F_{n,n}(r_1) \leqslant \liminf_{n \to \infty} F_{n,n}(t) \leqslant \limsup_{n \to \infty} F_{n,n}(t) \leqslant \limsup_{n \to \infty} F_{n,n}(r_2),$$

thus,

$$F_u(r_1) \leqslant \liminf_{n \to \infty} F_{n,n}(t) \leqslant \limsup_{n \to \infty} F_{n,n}(t) \leqslant F_u(r_2).$$

Therefore, if $r_1, r_2 \to t$, then $\lim_n F_{n,n}(t) = F_u(t)$.

Step 4 Now from $(F_{n,n})_{n \in \mathbb{N}}$ we will choose a subsequence $(F_{n,n,n})_{n \in \mathbb{N}}$ convergent to some distribution function F at every continuity point of F. To do this, let us choose two decreasing sequences $\varepsilon_1 = \varepsilon > \varepsilon_2 > \ldots$, $\lim_n \varepsilon_n = 0$, and $u = u_1 > u_2 > \ldots$, $\lim_{n \to \infty} u_n = 0$ which satisfy the following condition: for every fixed k, there exists a number $n_k \in \mathbb{N}$ such that for $n \geqslant n_k$, we have

$$F_n(2/u_k) - F_n(-2/u_k) \geqslant 1 - \varepsilon_k.$$

The existence of such sequences follows from step 1. Let S_{u_k} denote the set of all rational numbers in the set $I_{u_k} = (-2/u_k, 2/u_k)$. Of course, $S_{u_1} \subset S_{u_2} \subset \ldots$ Now,

6.3 Weak Convergence of Distributions

using the method described in step 2, from the sequence $(F_{n,n})_{n\in\mathbb{N}} = (F_{n,n,1})_{n\in\mathbb{N}}$ we choose a subsequence $(F_{n,n,2})_{n\in\mathbb{N}}$ convergent at every point $s \in S_{u_2} \setminus S_{u_1}$. Thus,

$$\lim_{n\to\infty} F_{n,n,2}(s) = F_{u_2}(s) \text{ for every } s \in S_{u_2}.$$

As in step 3, we define the function F_{u_2}. The sequence $F_{n,n,2}(t)$ converges to $F_{u_2}(t)$ at each continuity point $t \in I_{u_2}$ of the distribution function F_{u_2}. It is easy to see that $F_{u_2}(t) = F_{u_1}(t)$ for every $t \in I_{u_1}$.

Now, from the sequence $F_{n,n,2}$, we choose a subsequence $F_{n,n,3}$ convergent at each point $s \in S_{u_3} \setminus S_{u_2}$ and define the distribution function F_{u_3} such that $\lim_n F_{n,n,3}(t) = F_{u_3}(t)$ at each continuity point $t \in I_{u_3}$ of the distribution function F_{u_3}. It follows from the construction that for every $t \in I_{u_2}$, $F_{u_3}(t) = F_{u_2}(t)$.

Continuing, we obtain the infinite table of distribution functions $(F_{n,n,k})$, $n, k \in \mathbb{N}$ in which each row is a subsequence of the sequence in the previous row. We also get a sequence of distribution functions F_{u_k}, $k \in \mathbb{N}$, such that $\lim_n F_{n,n,k}(t) = F_{u_k}(t)$ at each continuity point $t \in I_{u_k}$ of the distribution function F_{u_k}. Moreover, for $j < k$,

$$F_{u_k}(t) = F_{u_j}(t) \text{ for every } t \in I_{u_j}.$$

Let now

$$F(t) = \lim_{k\to\infty} F_{u_k}(t).$$

The function F is well defined because the sequences $F_{u_k}(t)$, $t \in \mathbb{R}$, are constant beyond a finite number of elements. This construction shows that F is nondecreasing, left-continuous, and takes values in the interval $[0, 1]$. Since for $n \geq n_k$ we have $F_n(2/u_k) - F_n(-2/u_k) \geq 1 - \varepsilon_k$, it follows that $F(\infty) - F(-\infty) \geq 1$ and, therefore, we conclude that F is a distribution function.

If $t \in \mathbb{R}$ is a continuity point for the distribution function F, then there exists a number $k \in \mathbb{N}$ such that $t \in I_{u_k}$. Now, we have

$$F(t) = F_{u_k}(t) = \lim_{n\to\infty} F_{n,n,k}(t) = \lim_{n\to\infty} F_{n,n,n}(t),$$

where the last equality follows from the fact that the sequence $(F_{n,n,n})_{n\in\mathbb{N}}$ is a subsequence of $(F_{n,n,k})_{n\in\mathbb{N}}$.

Step 5 Let $(X_{n,n,n})$ be the subsequence of (X_n) that corresponds to the sequence of distribution functions $(F_{n,n,n})$. The carried out construction shows that $(X_{n,n,n})$ weakly converges to a random variable with distribution function F. By virtue of Remark 6.19, we get that

$$\int_{-\infty}^{\infty} e^{itx}\, dF(x) = \lim_{n\to\infty} \int_{-\infty}^{\infty} e^{itx}\, dF_{n,n,n}(x) = \lim_{n\to\infty} \int_{-\infty}^{\infty} e^{itx}\, dF_n(x) = \varphi(t).$$

It remains to prove that $X_n \Rightarrow X$, i.e., that the whole sequence converges weakly to the variable X. Assume that this is not the case. Then, for some continuous and bounded function f and some subsequence (n_k), we would have $|\int f \, dF_{n_k} - \int f \, dF| > \delta > 0$. Thus, X_{n_k} would not contain a subsequence weakly convergent to X. However, since $\lim \varphi_{n_k} = \varphi$, by virtue of the construction presented in steps 1 to 4, there must exist a subsequence (n_{k_l}) of (n_k) for which $X_{n_{k_l}} \Rightarrow X$, which contradicts our assumption. $\square$

6.3.1 Exercises

322. Assume that $X_n \xrightarrow{d} X$. Is it true that $X_n - X \xrightarrow{d} 0$?
323. Prove that if $X_n \xrightarrow{d} X$ and $a, b \in \mathbb{R}$, then $aX_n + b \xrightarrow{d} aX + b$.
324. **Poisson's Law of Small Numbers.** Assume that (X_n) is a sequence of random variables with Bernoulli distributions $B(n, p_n)$, respectively, and $\lim_{n\to\infty} np_n = \lambda > 0$. Prove that

$$\lim_{n\to\infty} \mathbf{P}\{X_n = k\} = \lim_{n\to\infty} \binom{n}{k} p_n^k (1-p_n)^{n-k} = \frac{\lambda^k}{k!} e^{-\lambda}, \quad k = 0, 1, \ldots$$

325. Let (X_n) be a sequence of symmetric random variables with identical Cauchy distributions, and let (a_n) be a sequence of positive real numbers such that $\sum_{k=1}^{\infty} a_k = A < \infty$. Does the sequence $Y_n = \sum_{k=1}^{n} a_k X_k$ converge in distribution?
326. Prove that if X_n have Poisson distributions with parameters λ_n, respectively, and $\lambda_n \to \lambda > 0$, then $X_n \xrightarrow{d} X$, for some random variable having the Poisson distribution with parameter λ.
327. Let $\Omega = [0, 1]$ and let $\mathbf{P}$ be the Lebesgue measure on Ω. For each $n \in \mathbb{N}$, we divide $[0, 1]$ into n equal parts and define the variable X_n by the following formula:

$$X_n(\omega) = \sum_{k=1}^{n} x_{k,n} \mathbf{1}_{\left(\frac{k-1}{n}, \frac{k}{n}\right]}(\omega),$$

where $x_{k,n}$ is any point selected from the interval $\left(\frac{k-1}{n}, \frac{k}{n}\right]$. Show that the sequence (X_n) converges in distribution to a variable with uniform distribution on the interval $[0, 1]$.
328. We say that the family of distributions $\{\mu_n : n \in \mathbb{N}\}$ is *tight* if for every $\varepsilon > 0$ there exists a compact set K such that $\mu_n(K) > 1 - \varepsilon$ for every $n \in \mathbb{N}$.
Let $\varphi, \varphi_1, \varphi_2, \ldots$ be characteristic functions of distributions $\mu, \mu_1, \ldots$ Prove that if $\lim \varphi_n(t) = \varphi(t)$ for every $t \in \mathbb{R}$, then the family of distributions $\{\mu_n : n \in \mathbb{N}\}$ is tight.

329. Prove that $(\mu_n)_{n\in\mathbb{N}}$ is weakly convergent if and only if each of its subsequences contains a weakly convergent subsequence.

6.4 Characteristic Functions of Random Vectors

Definition 6.20 The function $\varphi = \varphi_\mathbf{X}: \mathbb{R}^n \to \mathbb{C}$ is a *characteristic function of the random vector* $\mathbf{X} = (X_1, \ldots, X_n)$ if

$$\forall \overline{\xi} = (\xi_1, \ldots, \xi_n) \in \mathbb{R}^n \qquad \varphi_\mathbf{X}(\overline{\xi}) = \mathbf{E}e^{i\langle \mathbf{X}, \overline{\xi}\rangle} = \mathbf{E}\exp\left\{i\sum_{k=1}^n \xi_k X_k\right\}.$$

The notation $\langle \overline{\xi}, \mathbf{x}\rangle = \sum_{k=1}^n \xi_k x_k$ is understood as the inner product of two (row) vectors. However, here, it is more convenient to treat it as a product of two matrices: $\overline{\xi}$ of dimension $1 \times n$ and $\mathbf{x}^T$ of dimension $n \times 1$, i.e., $\langle \overline{\xi}, \mathbf{x}\rangle = \overline{\xi}\mathbf{x}^T$.

The properties of multidimensional characteristic functions are analogous to those of characteristic functions of random variables, and they can be proved in a similar way. In particular, for any $\overline{\xi}, \mathbf{m} \in \mathbb{R}^n$, we have that

$$\varphi_{\mathbf{X}-\mathbf{m}}(\overline{\xi}) = \varphi_\mathbf{X}(\overline{\xi})e^{i\langle \overline{\xi}, \mathbf{m}\rangle}.$$

Note that

$$\varphi_{(X_1,\ldots,X_n)}(\xi, 0, \ldots, 0) = \mathbf{E}e^{i\xi X_1} = \varphi_{X_1}(\xi),$$

so it is a characteristic function of the variable X_1. Moreover,

$$\varphi_{(X_1,\ldots,X_n)}(\xi_1 t, \ldots, \xi_n t) = e^{it\sum_{k}^n \xi_k X_k} = \varphi_{\sum_k^n \xi_k X_k}(t).$$

Note also that for any matrix A of dimension $n \times n$ (corresponding to the linear operator $T_A : \mathbf{x} \to \mathbf{x}A$), we have

$$\langle \overline{\xi}, \mathbf{x}A\rangle = \overline{\xi}(\mathbf{x}A)^T = (\overline{\xi}A^T)\mathbf{x}^T.$$

Hence,

$$\varphi_{\mathbf{X}A}(\overline{\xi}) = \varphi_\mathbf{X}(\overline{\xi}A^T).$$

If the random vector $\mathbf{Y} = (Y_1, \ldots, Y_n)$ is such that the random variables $Y_1, \ldots, Y_n$ are independent with the standard normal distribution, then

$$\varphi_\mathbf{Y}(\overline{\xi}) = \exp\left\{-\frac{1}{2}\sum_{k=1}^n \xi_k^2\right\} = \exp\left\{-\frac{1}{2}\overline{\xi}\,\overline{\xi}^T\right\}.$$

Therefore for any matrix A of the dimension $n \times n$, we have

$$\varphi_{YA}(\bar{\xi}) = \exp\left\{-\frac{1}{2}(\bar{\xi}A^T)(\bar{\xi}A^T)^T\right\} = \exp\left\{-\frac{1}{2}\bar{\xi}\left(A^T A\right)\bar{\xi}^T\right\}.$$

We did not assume here that the matrix A is non-degenerate, which means that the above formula is valid even when $T_A(\mathbb{R}^n) \subsetneq \mathbb{R}^n$.

The uniqueness theorem for a multidimensional characteristic function takes the following form:

Theorem 6.21 *If $\varphi : \mathbb{R}^n \to \mathbb{C}$ is a characteristic function of the random vector $\mathbf{X} \in \mathbb{R}^n$ with distribution μ, and if $B = [a_1, b_1] \times \cdots \times [a_n, b_n]$ is a rectangle in $\mathbb{R}^n$ such that $\mu(\partial B) = 0$, where ∂B is the boundary of the set B, then*

$$\mu(B) = \lim_{T \to \infty} \frac{1}{(2\pi)^n} \int_{[-T,T]^n} \cdots \int \prod_{k=1}^{n} \frac{e^{-ia_k t_k} - e^{-ib_k t_k}}{it_k} \varphi(t_1, \ldots, t_n)\, dt_1 \ldots dt_n.$$

If φ is a function on $\mathbb{R}^n$ which is integrable in absolute value, then μ has a density function given by

$$f(\mathbf{x}) = \frac{1}{(2\pi)^n} \int_{\mathbb{R}^n} \cdots \int e^{-i\sum^n x_k t_k} \varphi(t_1 \ldots t_n)\, dt_1 \ldots dt_n.$$

From the uniqueness theorem, it is easy to derive the following independence criterion for random variables:

Theorem 6.22 *Random variables $X_1, \ldots, X_n$ are independent if and only if*

$$\varphi_{(X_1,\ldots,X_n)}(t_1, \ldots, t_n) = \varphi_{X_1}(t_1) \ldots \varphi_{X_n}(t_n).$$

The next theorem reduces the study of convergence in distribution for a sequence of random vectors to the study of convergence in distribution of the corresponding random variables.

Theorem 6.23 (Cramér–Wold Theorem) *If $\mathbf{X}_1, \mathbf{X}_2, \ldots$ are random vectors taking values in $\mathbb{R}^n$, then $\mathbf{X}_k \xrightarrow{d} \mathbf{X}$, for $k \to \infty$ if and only if for every $\bar{\xi} \in \mathbb{R}^n$*

$$\langle \bar{\xi}, \mathbf{X}_k \rangle \xrightarrow{d} \langle \bar{\xi}, \mathbf{X} \rangle.$$

Thus, convergence in distribution is equivalent to convergence of the characteristic functions also in the case of random vectors.

6.4.1 Exercises

330. Assume that the function $\varphi(\xi_1, \ldots, \xi_n)$ is the characteristic function of a random vector $\mathbf{X} = (X_1, \ldots, X_n)$. Find the characteristic function of the random variable $Y = \sum_{k=1}^{n} a_k X_k$, where $a_1, \ldots, a_n \in \mathbb{R}$.
331. The random vector $\mathbf{X} = (X_1, \ldots, X_n)$ has a Gaussian distribution with covariance matrix Σ and vector of expected values $\mathbf{m} = (m_1, \ldots, m_n)$. Calculate the characteristic function of $\mathbf{X}$.
332. The function $\varphi(t, s) = \exp\{-|at + bs|\}$ is the characteristic function of a random vector (X, Y). Are the variables X, Y independent? Does the vector (X, Y) have an absolutely continuous distribution with respect to Lebesgue measure?
333. Let $\mathbf{X}_1, \mathbf{X}_2, \cdots \in \mathbb{R}^n$ be independent random vectors with the same multivariate normal distribution, expectation zero and covariance matrix Σ. Find the limit distribution for the series of random vectors
$$Z_n = \frac{\mathbf{X}_1 + \cdots + \mathbf{X}_n}{\sqrt{n}}.$$
334. Find the characteristic function of the random vector (U, V) which is uniformly distributed on the unit sphere $S_1 \subset \mathbb{R}^2$.
    Hint. This vector has the same distribution as $(\cos\theta, \sin\theta)$, where θ is uniformly distributed over $[0, 2\pi]$.
335. By calculating the corresponding characteristic functions show that the random variable $R := \sqrt{X^2 + Y^2}$ and the random vector $(U, V) := \left(\frac{X}{R}, \frac{Y}{R}\right)$ are independent if (X, Y) has a standard normal distribution on $\mathbb{R}^2$ with expectation zero and covariance matrix I_2.

Chapter 7
Limit Theorems

The Central Limit Theorem, the Law of Large Numbers and their numerous variants play a special role in Probability Theory. What they have in common is the consideration of various methods of describing the limit behavior of random variables

$$\frac{X_1 + \cdots + X_n - a_n}{b_n},$$

for some constants $a_n, b_n \in \mathbb{R}$, $b_n \neq 0$. To study such normalized sums, criteria for convergence of series of independent random variables are useful. As we will see in the next section, these sums are convergent exclusively only almost everywhere or almost nowhere.

7.1 Kolmogorov's Zero-One Law

Consider a sequence of σ-fields $(\mathcal{F}_n)_n$ which are independent, i.e., for every choice of different $n, k \in \mathbb{N}$ and any $A \in \mathcal{F}_n$, $B \in \mathcal{F}_k$, we have $\mathbf{P}(A \cap B) = \mathbf{P}(A)\mathbf{P}(B)$. For every $n \in \mathbb{N}$, we define

$$\mathcal{G}_n = \sigma(\mathcal{F}_1, \ldots, \mathcal{F}_n), \qquad \mathcal{F}_{n,\infty} = \sigma(\mathcal{F}_n, \mathcal{F}_{n+1}, \ldots).$$

By the *tail σ-field*, we understand the following:

$$\mathcal{F}_\infty = \bigcap_{n=1}^{\infty} \mathcal{F}_{n,\infty}.$$

The elements of the tail σ-field are called *tail events* or *rare events*. More plainly, we say that tail events are precisely those events whose occurrence can still be determined if an arbitrarily large but finite initial segment of the σ-fields $\mathcal{F}_k$ is removed. Now, we can formulate the following:

Theorem 7.1 (Kolmogorov's Zero-One Law) *Any tail event $A \in \mathcal{F}_\infty$ has probability either zero or one. We can also express it differently: a tail event will either almost surely happen or almost surely not happen.*

Proof Note that for every n, the σ-fields $\mathcal{G}_n$ and $\mathcal{F}_{n+1,\infty}$ are independent. Since $A \in \mathcal{F}_\infty$, for every $n \in \mathbb{N}$, we have $A \in \mathcal{F}_{n+1,\infty}$. Therefore, A is independent of each σ-field $\mathcal{G}_n$. Consequently, A is independent of $\sigma(\mathcal{F}_1, \mathcal{F}_2, \ldots) = \mathcal{F}_{1,\infty} \supset \mathcal{F}_\infty$. In particular, we can conclude that A is independent of itself, which ends the proof. □

Sometimes, it can be easy to apply Kolmogorov's zero-one law to show that some event has probability 0 or 1, but it is much harder to determine which value is correct. In such cases, we can use an elegant method and prove that the probability of the given tail event exceeds ε for some $\varepsilon > 0$. This guarantees that the probability of this event is one.

Example of Application 7.2 If (X_n) is a sequence of independent random variables, then the series $\sum_{n=1}^{\infty} X_n$ converges with probability zero or one.

Proof Let $\mathcal{F}_n = \sigma(X_n)$. Then, the σ-fields $\mathcal{F}_n$ are totally independent. Moreover, for each $n \in \mathbb{N}$, we have

$$\left\{\omega: \sum_{k=1}^{\infty} X_k(\omega) \text{ converges}\right\} = \left\{\omega: \sum_{k=n}^{\infty} X_k(\omega) \text{ converges}\right\} \in \mathcal{F}_{n,\infty}.$$

Now, all we need to do is to apply Kolmogorov's zero-one law. □

7.1.1 Exercises

336. Let (X_n) be a sequence of independent random variables, $\mathcal{F}_n = \sigma(X_n)$. Prove that the following events belong to $\mathcal{F}_\infty$,

$$\left\{\text{there exists a finite limit } \lim_{n\to\infty} X_n\right\},$$

$$\left\{X_n = \infty\right\}, \left\{\lim_{n\to\infty} \frac{X_1 + \cdots + X_n}{n} \leqslant a\right\}.$$

337. Let (X_n) be a sequence of independent random variables, $\mathcal{F}_n = \sigma(X_n)$. Prove that if the random variable X is $\mathcal{F}_\infty$-measurable, then it is constant.

338. Let (X_n) be a sequence of independent random variables, $\mathcal{F}_n = \sigma(X_n)$. Prove that the radius of convergence of the power series $\sum_{n=1}^{\infty} X_n x^n$ (for x real or complex) is constant almost everywhere.

7.2 Laws of Large Numbers

To start with, we will discuss a series of theorems known as the weak laws of large numbers. Why weak? This name is related to the type of convergence of sequences of random variables appearing in these theorems, the type called *convergence in probability*.

Definition 7.3 We say that a sequence of random variables (X_n) defined on the same probabilistic space $(\Omega, \mathcal{F}, \mathbf{P})$ *converges in probability* to the random variable X (notation: $X_n \xrightarrow{P} X$) if for every $\varepsilon > 0$

$$\lim_{n \to \infty} \mathbf{P}\{\omega : |X_n(\omega) - X(\omega)| < \varepsilon\} = 1.$$

Example 7.4 To understand why the convergence just defined is not very strong, consider the following example: $\Omega = [0, 1]$, $\mathbf{P}$ is the normalized Lebesgue measure on Ω, and the sequence of random variables X_n is defined as follows:

$$X_n(\omega) = \mathbf{1}_{(0, \frac{1}{n})}(\{n\pi + \omega\}),$$

where $\{a\}$ denotes the fractional part of a number a. It is easy to see that X_n is equal to 1 on a set of length $\frac{1}{n}$, and on the remaining part of the interval $[0, 1]$, it equals zero. Therefore, we get:

$$\mathbf{P}\{\omega : |X_n - 0| < \varepsilon\} \geq 1 - \frac{1}{n} \xrightarrow{n \to \infty} 1.$$

It follows that $X_n \xrightarrow{P} 0$. However, if we set $\omega \in \Omega$, we know that for any $n \in \mathbb{N}$, there exist $k, \ell > n$ for which

$$\{k\pi + \omega\} \in \left(0, \tfrac{1}{k}\right), \text{ thus } X_k(\omega) = 1;$$
$$\{k\pi + \omega\} \notin \left(0, \tfrac{1}{k}\right), \text{ thus } X_k(\omega) = 0.$$

This means that $(X_n(\omega))_{n \in \mathbb{N}}$ is a sequence containing infinitely many ones and infinitely many zeros, which implies that it is a non-convergent sequence. It is known, however, that for sufficiently large values of n, the probability of the event that $X_n(\omega) = 0$ is close to one.

Although probability convergence is "weak", the limit of a sequence of random variables convergent in probability is uniquely determined, as can be seen from the following lemma.

Lemma 7.5 *If $X_n \xrightarrow{P} X$ and $X_n \xrightarrow{P} Y$, then $\mathbf{P}\{X = Y\} = 1$.*

Proof From the assumptions of this lemma, it follows that:

$$\mathbf{P}\{|X - Y| \geq \varepsilon\} \leq \mathbf{P}\{|X - X_n| \geq \varepsilon/2 \text{ or } |X_n - Y| \geq \varepsilon/2\}$$
$$\leq \mathbf{P}\{|X - X_n| \geq \varepsilon/2\} + \mathbf{P}\{|X_n - Y| \geq \varepsilon/2\} \xrightarrow{n \to \infty} 0.$$

Hence, we have

$$\mathbf{P}\{X \neq Y\} = \mathbf{P}\left(\bigcup_{k=1}^{\infty}\{|X - Y| \geq 1/k\}\right) = \lim_{k \to \infty} \mathbf{P}\{|X - Y| \geq 1/k\} = 0,$$

which was to be shown. □

Theorem 7.6 (Markov's Weak Law of Large Numbers) *If (X_n) is a sequence of random variables such that*

$$\lim_{n \to \infty} \frac{1}{n^2} \mathrm{Var}\left(\sum_{k=1}^{n} X_k\right) = 0,$$

then for every $\varepsilon > 0$, we have

$$\lim_{n \to \infty} \mathbf{P}\left\{\omega : \left|\frac{1}{n}\sum_{k=1}^{n}(X_k - \mathbf{E}X_k)\right| < \varepsilon\right\} = 1.$$

Proof Let $Y_n = \frac{1}{n}\sum_{k=1}^{n} X_k$. It follows from our assumptions that

$$\lim_{n \to \infty} \mathrm{Var}(Y_n) = 0.$$

By virtue of Chebyshev's inequality, we get:

$$\mathbf{P}\left\{\omega : \left|\frac{1}{n}\sum_{k=1}^{n}(X_k - \mathbf{E}X_k)\right| < \varepsilon\right\} = \mathbf{P}\{\omega : |Y_n - \mathbf{E}Y_n| < \varepsilon\}$$
$$= 1 - \mathbf{P}\{\omega : |Y_n - \mathbf{E}Y_n| \geq \varepsilon\} \geq 1 - \frac{1}{\varepsilon^2}\mathrm{Var}(Y_n) \xrightarrow{n \to \infty} 1,$$

which ends the proof. □

7.2 Laws of Large Numbers

The proven theorem, as well as the Bernoulli and Chebyshev Theorems described in the exercises, requires the existence of finite variances of the variables under consideration. Here, we present another form of the Weak Law of Large Numbers in which we abandon this assumption at the expense of the assumption of identical distributions of random variables. We make this assertion without proof. The next theorem, Chinczyn's Law of Large Numbers, is presented without proof since it follows trivially from Lemma 7.9 and Kolmogorov's Second Strong Law of Large Numbers (to be described in Theorem 7.13 further in this section).

Theorem 7.7 (Chinczyn's Law of Large Numbers) *If $X_1, X_2, \ldots$ is a sequence of independent random variables with identical distributions and finite expectation $m = \mathbf{E}X_i$, then for every $\varepsilon > 0$*

$$\lim_{n \to \infty} \mathbf{P}\left\{\omega : \left|\frac{1}{n}\sum_{k=1}^{n} X_k - m\right| < \varepsilon\right\} = 1.$$

Another type of convergence of random variables is *convergence with probability 1*, which we also call *convergence almost everywhere* or *almost sure convergence*.

Definition 7.8 We say that the sequence of random variables (X_n) defined on the same probability space $(\Omega, \mathcal{F}, \mathbf{P})$ *almost surely converges* to the random variable X (notation $X_n \to X$ a.e.) if

$$\mathbf{P}\left\{\omega : \lim_{n \to \infty} X_n(\omega) = X(\omega)\right\} = 1.$$

Note that convergence in probability does not imply almost sure convergence. For the random variables described in Example 7.4, which converge in probability to $X \equiv 0$, we get:

$$\mathbf{P}\left\{\omega : \text{the limit } \lim_{n \to \infty} X_n(\omega) \text{ exists}\right\} = \mathbf{P}\{\emptyset\} = 0.$$

This means that the sequence of random variables (X_n) is almost surely divergent. The next lemma shows that the opposite implication holds, so convergence with probability 1 is actually "stronger" than convergence in probability.

Lemma 7.9 *Let (X_n) be a sequence of random variables defined in the same probability space $(\Omega, \mathcal{F}, \mathbf{P})$. Then,*

$$X_n \to X \text{ a.e.} \implies X_n \xrightarrow{P} X.$$

Proof Assume that $X_n \to X$ a.e. This means that

$$\mathbf{P}\left\{\omega : X_n(\omega) \to X(\omega)\right\} = 1.$$

Equivalently, we can write it as

$$1 = \mathbf{P}\Big\{\omega : \forall \varepsilon > 0, \exists n(\omega,\varepsilon) \forall k > n(\omega,\varepsilon) \big| X_k(\omega) - X(\omega) \big| < \varepsilon \Big\}$$

$$= \mathbf{P}\bigg(\bigcap_{\varepsilon>0} \bigcup_{n\in\mathbb{N}} \bigcap_{k \geq n} \{\omega : |X_k(\omega) - X(\omega)| < \varepsilon\}\bigg).$$

This means that for any $\varepsilon > 0$,

$$1 = \mathbf{P}\bigg(\bigcup_{n\in\mathbb{N}} \bigcap_{k \geq n} \{\omega : |X_k(\omega) - X(\omega)| < \varepsilon\}\bigg),$$

or, equivalently,

$$0 = \mathbf{P}\bigg(\bigcap_{n\in\mathbb{N}} \bigcup_{k \geq n} \{\omega : |X_k(\omega) - X(\omega)| \geq \varepsilon\}\bigg).$$

Since the sequence of sets $\bigcup_{k \geq n} \{\omega : |X_k(\omega) - X(\omega)| \geq \varepsilon\}$ is decreasing, we have

$$0 = \lim_{n\to\infty} \mathbf{P}\bigg(\bigcup_{k \geq n} \{\omega : |X_k(\omega) - X(\omega)| \geq \varepsilon\}\bigg).$$

To conclude the proof, it is enough to note now that

$$\{|X_n - X| > \varepsilon\} \subset \bigcup_{k=n}^{\infty} \{|X_k - X| > \varepsilon\}.$$

□

Now, we will give two Kolmogorov theorems, in other words, two versions of the Strong Law of Large Numbers. We will begin by recalling two lemmas from analysis that we will need to ensure completeness of the proofs. The first one generalizes the well-known fact that if a sequence converges, then the sequence of arithmetic means converges to the same limit.

Lemma 7.10 (Toeplitz's Lemma) *Let (a_n) be a sequence of non-negative numbers, $b_n = \sum_{k=1}^{n} a_k$, $b_n > 0$ for all $n \in \mathbb{N}$, $b_n \nearrow \infty$. If (x_n) is a sequence such that $\lim_{n\to\infty} x_n = x \in \mathbb{R}$, then*

$$\lim_{n\to\infty} \frac{1}{b_n} \sum_{k=1}^{n} a_k x_k = x.$$

7.2 Laws of Large Numbers

Proof Let $\varepsilon > 0$. We can choose $n_0 \in \mathbb{N}$ such that for $n \geq n_0$, we have the inequality: $|x_n - x| < \varepsilon/2$. Since $b_n \nearrow \infty$, there also exists an $n_1 > n_0$ such that

$$\frac{1}{b_{n_1}} \sum_{k=1}^{n_0} a_k |x_k - x| < \frac{\varepsilon}{2}.$$

Finally, for $n > n_1$, we have

$$\left| \frac{1}{b_n} \sum_{k=1}^{n} a_k x_k - x \right| \leq \frac{1}{b_n} \sum_{k=1}^{n_0} a_k |x_k - x| + \frac{1}{b_n} \sum_{k=n_0+1}^{n_0} a_k |x_k - x|$$

$$< \frac{b_1}{b_n} \frac{\varepsilon}{2} + \frac{\varepsilon}{2} \frac{1}{b_n} \sum_{k=n_0+1}^{n_0} a_k < \varepsilon.$$

$\square$

Lemma 7.11 (Kronecker's Lemma) *Let (b_n) be an increasing sequence of positive numbers, $\lim_n b_n = \infty$, and let (x_n) be a sequence of real numbers such that $\sum_{n=1}^{\infty} x_n$ is convergent. Then,*

$$\lim_{n \to \infty} \frac{1}{b_n} \sum_{k=1}^{n} b_k x_k = 0.$$

Proof Let $b_0 = 0$, $s_0 = 0$, $s_n = \sum_{k=1}^{n} x_n$. Then,

$$\sum_{k=1}^{n} b_k x_k = \sum_{k=1}^{n} b_k (s_k - s_{k-1}) = b_n s_n - b_0 s_0 - \sum_{k=1}^{n} s_{k-1}(b_k - b_{k-1}).$$

Since the limit $\lim_{n \to \infty} s_n$ exists, by Toeplitz's Lemma we obtain that

$$\frac{1}{b_n} \sum_{k=1}^{n} b_k x_k = s_n - \frac{1}{b_n} \sum_{k=1}^{n} s_{k-1}(b_k - b_{k-1}) \longrightarrow 0 \quad \text{for } n \to \infty.$$

$\square$

In particular, for $b_n = n$, $x_n = n^{-1} y_n$, we have the following implication:

$$\sum_{n=1}^{\infty} \frac{y_n}{n} \text{ is convergent} \quad \Longrightarrow \quad \lim_{n \to \infty} \frac{y_1 + \cdots + y_n}{n} = 0.$$

Theorem 7.12 (Kolmogorov's First Strong Law of Large Numbers) *If $X_1, X_2, \ldots$ is a sequence of random variables with finite variances and*

$$\sum_{n=1}^{\infty} \frac{1}{n^2} \operatorname{Var} X_n < \infty,$$

then

$$\mathbf{P}\left\{\omega : \lim_{n\to\infty} \frac{1}{n} \sum_{k=1}^{n} (X_k - \mathbf{E}X_k) = 0\right\} = 1.$$

Proof Let $S_n = \sum_{k=1}^{n} (X_k - \mathbf{E}X_k)$. Since

$$\frac{S_n - \mathbf{E}S_n}{n} = \frac{1}{n} \sum_{k=1}^{n} k \cdot \frac{X_k - \mathbf{E}X_k}{k},$$

then, by Kronecker's Lemma, it is enough to prove that $\sum_{k=1}^{\infty} \frac{1}{k}(X_k - \mathbf{E}X_k)$ converges almost everywhere. To see this, we shall check that the Cauchy condition holds. From Chebyshev's inequality and the independence of $X_k, k \in \mathbb{N}$, we have that for every $\varepsilon > 0$:

$$\mathbf{P}\left\{\sup_{k,l \geqslant n} |S_k - S_l| > \varepsilon\right\} \leqslant \mathbf{P}\left\{2 \sup_{k \geqslant n} |S_k - S_l| > \varepsilon\right\}$$

$$= \lim_{m \to \infty} \mathbf{P}\left\{\sup_{n \leqslant k \leqslant m} |S_k - S_n| > \varepsilon/2\right\}$$

$$\leqslant \lim_{m \to \infty} 3 \sup_{n \leqslant k \leqslant m} \mathbf{P}\left\{|S_k - S_n| > \varepsilon/6\right\}$$

$$\leqslant \lim_{m \to \infty} \frac{108}{\varepsilon^2} \sum_{j=n}^{m} \mathbf{E}(X_j - \mathbf{E}X_j)^2.$$

The second inequality here follows from the Lévy–Ottaviani inequality:

$$\mathbf{P}\left\{\max_{i \leqslant n} |S_i| > \varepsilon\right\} \leqslant 3 \max_{i \leqslant n} \mathbf{P}\left\{|S_i| > \varepsilon/3\right\},$$

which holds for sums of independent random variables. The right side of this inequality tends to zero when $n \to \infty$ as the remainder of a convergent series. Thus, S_n converges almost everywhere. □

Theorem 7.13 (Kolmogorov's Second Strong Law of Large Numbers) *If $X_1, X_2, \ldots$ is a sequence of independent random variables with identical*

7.2 Laws of Large Numbers

distributions and finite expectation $m = \mathbf{E}X_k$, *then*

$$\mathbf{P}\left\{\omega : \lim_{n\to\infty} \frac{1}{n} \sum_{k=1}^{n} X_k = m\right\} = 1.$$

Proof Since the convergence of $\frac{1}{n}\sum^n X_k \to \mathbf{E}X_1$ is equivalent to the convergence of $\frac{1}{n}\sum^n (X_k - \mathbf{E}X_1) \to 0$, we can assume without loss of generality that $\mathbf{E}X_1 = 0$. It follows from Theorem 4.15 that

$$\sum_{n=1}^{\infty} \mathbf{P}\{|X_1| \geq n\} \leq \mathbf{E}|X_1| \leq 1 + \sum_{n=1}^{\infty} \mathbf{P}\{|X_1| \geq n\},$$

hence $\sum_{n=1}^{\infty} \mathbf{P}\{|X_1| \geq n\} < \infty$. Let $Y_n = X_n \mathbf{1}_{\{|X_n|\geq n\}}$, i.e., Y_n is the truncation of X_n at the level n. Since the variables X_n have identical distributions, we have

$$\sum_{n=1}^{\infty} \mathbf{P}\{X_n \neq X_n\} = \sum_{n=1}^{\infty} \mathbf{P}\{|X_n| > n\} = \sum_{n=1}^{\infty} \mathbf{P}\{|X_1| > n\} \leq \mathbf{E}|X_1| < \infty.$$

From the Borel–Cantelli Lemma, it follows that

$$\lim_{n\to\infty} \frac{X_1 + \cdots + X_n}{n} \to 0\, a.e. \quad \text{if and only if} \quad \lim_{n\to\infty} \frac{Y_1 + \cdots + Y_n}{n} \to 0\, a.e.$$

Since $\mathbf{E}Y_n = \mathbf{E}(X_1 \mathbf{1}_{\{|X_n|\leq n\}}) \to \mathbf{E}X_1 = 0$ (the simplifying assumption), it is enough to show that

$$\lim_{n\to\infty} \frac{(Y_1 - \mathbf{E}Y_1) + \cdots + (Y_n - \mathbf{E}Y_n)}{n} = 0\, a.e.$$

To do this, we shall use Kolmogorov's First Strong Law of Large Numbers, which shows that $\sum_{n=1}^{\infty} \frac{1}{n^2} \mathrm{Var} Y_n < \infty$. First, note that

$$|Y_n - \mathbf{E}Y_n| = |X_n - \mathbf{E}Y_n|\mathbf{1}_{\{|X_n|\leq n\}} + |\mathbf{E}Y_n|\mathbf{1}_{\{|X_n|\leq n\}}$$
$$\leq |X_n|\mathbf{1}_{\{|X_n|\leq n\}} + |\mathbf{E}Y_n|.$$

Since $|\mathbf{E}Y_n| \to 0$, we have $\sum_{n=1}^{\infty} |\mathbf{E}Y_n|^2 n^{-2} =: A < \infty$, and so

$$\sum_{n=1}^{\infty} \frac{\mathrm{Var} Y_n}{n^2} \leq \sum_{n=1}^{\infty} \frac{1}{n^2} \mathbf{E}\left(X_n^2 \mathbf{1}_{\{|X_n|\leq n\}}\right) + \sum_{n=1}^{\infty} \frac{|\mathbf{E}Y_n|^2}{n^2}$$
$$= \sum_{n=1}^{\infty} \frac{1}{n^2} \sum_{k=1}^{n} \mathbf{E}\left(X_1^2 \mathbf{1}_{\{k-1 < |X_1| \leq k\}}\right) + A$$

$$= \sum_{k=1}^{\infty} \mathbf{E}\left(X_1^2 \mathbf{1}_{\{k-1<|X_1|\leq k\}}\right) \sum_{n=k}^{\infty} \frac{1}{n^2} + A$$

$$\leq 2\sum_{k=1}^{\infty} \frac{1}{k} \mathbf{E}\left(X_1^2 \mathbf{1}_{\{k-1<|X_1|\leq k\}}\right) + A$$

$$\leq 2\sum_{k=1}^{\infty} \mathbf{E}\left(|X_1| \mathbf{1}_{\{k-1<|X_1|\leq k\}}\right) + A = 2\mathbf{E}|X_1| + A < \infty,$$

which was to be shown. □

7.2.1 Exercises

339. **Bernoulli's Weak Law of Large Numbers.** Prove that for the sequence of random variables S_n counting the number of successes in n Bernoulli trials with success probability p and for every $\varepsilon > 0$, we have

$$\lim_{n\to\infty} \mathbf{P}\left\{\omega : \left|\frac{1}{n}S_n - p\right| < \varepsilon\right\} = 1.$$

340. **Chebyshev's Weak Law of Large Numbers.** Prove that that if random variables $X_1, X_2, \ldots$ are independent and their variances are jointly bounded, then for every $\varepsilon > 0$,

$$\lim_{n\to\infty} \mathbf{P}\left\{\omega : \left|\frac{1}{n}\sum_{k=1}^{n}(X_k - \mathbf{E}X_k)\right| < \varepsilon\right\} = 1.$$

341. Check whether the Strong or Weak Law of Large Numbers holds for sequences of independent random variables (X_n) with the following distributions:
    (a) $\mathbf{P}\{X_n = 2^n\} = \mathbf{P}\{X_n = -2^n\} = 0.5$;
    (b) $\mathbf{P}\{X_n = 2^n\} = \mathbf{P}\{X_n = -2^n\} = 2^{-2n-1}$, $\mathbf{P}\{X_n = 0\} = 1 - 2^{-2n}$;
    (c) $\mathbf{P}\{X_n = n\} = \mathbf{P}\{X_n = -n\} = 0.5$;
    (d) $\mathbf{P}\{X_n = n\} = \mathbf{P}\{X_n = -n\} = 0.5n^{-1/2}$, $\mathbf{P}\{X_n = 0\} = 1 - n^{-1/2}$;
    (e) X_n has normal distribution $N(0, \sqrt{n})$;
    (f) X_n has Poisson distribution with parameter $\lambda = 2^{-n}$.

342. Random variables $X_1, X_2, \ldots$ are independent and $\mathbf{P}\{X_n = \pm\sqrt{\ln n}\} = \frac{1}{2}$. Does the Law of Large Numbers hold for the sequence (X_n)?

343. Random variables $X_1, X_2, \ldots$ are independent and $\mathbf{P}\{X_n = \pm n\alpha_n\} = \frac{1}{2}$, where $\alpha_n > \alpha$ for every $n \in \mathbb{N}$ and some $\alpha > 0$. Does the Law of Large Numbers hold for this sequence?

344. Let $X_1, X_2, \ldots$ be a sequence of 1-dependent random variables, i.e., the random variable X_n may depend on X_{n-1} and X_{n+1}, but it is independent of the other variables. Prove that if

$$\lim_{n \to \infty} n^{-1} \text{Var} X_n = 0,$$

then for the sequence (X_n), the Weak Law of Large Numbers holds.

7.3 The Central Limit Theorem

Theorem 7.14 (Lindeberg–Lévy Central Limit Theorem) *Let $X_1, X_2, \ldots$ be a sequence of independent identically distributed random variables with parameters $\text{E} X_i = m$, $\text{Var} X_i = \sigma^2 < \infty$. Then for any real number t,*

$$\lim_{n \to \infty} \mathbf{P}\left\{ \frac{\sum_{k=1}^n X_k - nm}{\sigma \sqrt{n}} < t \right\} = \Phi(t),$$

where Φ is the distribution function for the normal distribution $N(0, 1)$.

Proof Without losing generality, we can assume that for every $i \in \mathbb{N}$, X_i has expected value equal to zero. It is enough to consider $X'_i = X_i - m$. Let $\varphi(t) = \text{E} e^{itX_1}$. By Theorem 6.11, we have $\varphi'(0) = 0$ and $\varphi''(0) = -\sigma^2$. Using the second order Taylor expansion for the function φ, we get that

$$\varphi(t) = 1 - \frac{\sigma^2 t^2}{2} + o(t^2),$$

where $o(x)/x \to 0$ as $x \to 0$. Using the independence and equality of distributions of variables X_i we conclude that, at each fixed point,

$$\mathbf{E} \exp\left\{ it \frac{\sum_{k=1}^n X_i}{\sigma \sqrt{n}} \right\} = (\varphi(t/\sigma \sqrt{n}))^n = \left(1 - \frac{t^2}{2n} + o(t^2/\sigma^2 n) \right)^n.$$

Now, it is not hard to see that the last expression tends to $\exp\{-t^2/2\}$ as $n \to \infty$. We have shown that the characteristic functions of the sequence of random variables $S_n/\sigma \sqrt{n}$ converge to the characteristic function of the standard normal distribution $N(0, 1)$. From the Lévy–Cramér Theorem 6.18, the convergence of the distributions (and thus the distribution functions of these distributions) to the limit distribution (limit distribution function) follows. □

The next theorem, proven by J.W. Lindeberg in 1922 as a generalization of many partial results, is presented here without proof.

Theorem 7.15 (Lindeberg–Feller Theorem) *Assume that the random variables $\{X_{n,k}: n \in \mathbb{N}\ k \in \{1,\ldots n\}\}$ with $m_{n,k} = \mathbf{E}X_{n,k}$, $\sigma_{n,k}^2 = \mathrm{Var}(X_{n,k})$ satisfy the following properties:*

(a) *for every $n \in \mathbb{N}$, the random variables $X_{n,1}, \ldots, X_{n,n}$ are independent and $\sum_{k=1}^n \sigma_{n,k}^2 \to 1$;*
(b) $\sum_{k=1}^n \mathbf{E}\big((X_{n,k} - m_{n,k})^2 \mathbf{1}_{|X_{n,k}-m_{n,k}|>\varepsilon}\big) \to 0$ *for all $\varepsilon > 0$.*

Then,
$$\frac{\sum_{k=1}^n (X_{n,k} - m_{n,k})}{\sqrt{n}} \xrightarrow{d} N(0, 1).$$

Condition (b) is known as the *Lindeberg condition*.

7.3.1 Exercises

345. **De Moivre–Laplace Theorem.** Using the Lindeberg–Lévy Theorem, prove that if (S_n) is a sequence of random variables with the Bernoulli distribution $S_n \sim B(n, p)$, then for any real number
$$\lim_{n \to \infty} \mathbf{P}\left\{\omega : \frac{S_n(\omega) - np}{\sqrt{npq}} < a\right\} = \Phi(a).$$

346. Random variables $X_1, \ldots, X_{100}$ are independent with the same Poisson distribution with parameter $\lambda = 2$. Find the approximate value of the expression
$$\mathbf{P}\left\{\omega : \sum_{k=1}^{100} X_k > 190\right\}.$$

347. There is a newspaper vendor in the street. Suppose that every passer-by buys a newspaper with probability $\frac{1}{3}$. Let X be the number of passers-by until the 100th paper is sold. Find the exact and asymptotic distribution of the random variable X.

348. A computer adds 1200 real numbers, each approximated to the nearest integer. We assume that the approximation errors are independent and uniformly distributed on the interval $[-\frac{1}{2}, \frac{1}{2}]$. Find the probability that the error in calculating this sum will exceed 10.

349. Let (X_n) be a sequence of independent random variables with equal distributions and finite variance, and let $Y_n = \sum_{k=1}^n X_k$. Prove that for any real numbers $a, b, a < b$
$$\lim_{n \to \infty} \mathbf{P}\{\omega : a < Y_n(\omega) < b\} = 0.$$

7.3 The Central Limit Theorem

350. Let (X_n) be a sequence of independent random variables with equal distributions and finite, non-zero variance. Prove that for any real number x, the limit

$$\lim_{n\to\infty} \mathbf{P}\{\omega : X_1(\omega) + \cdots + X_n(\omega) < x\}$$

exists and is equal to one of the three numbers: 0, 1, $\frac{1}{2}$. Identify the conditions under which each of these numbers appears.

351. Let (X_n) be a sequence of independent random variables with identical distributions, expected value zero, and finite variance. Prove that for any real number x, the limit

$$\lim_{n\to\infty} \mathbf{P}\left\{\omega : \left|\frac{X_1(\omega) + \cdots + X_n(\omega)}{n^\alpha}\right| \leqslant x\right\}$$

equals 0 if $\alpha \in (0, \frac{1}{2})$ and equals 1 if $\alpha > \frac{1}{2}$.

352. Let (X_n) be a sequence of independent random variables with the same uniform distributions on the interval $[0, 1]$ and let $Y_n = \sum_{k=1}^n X_k$. Find the sequence of real numbers (a_n) which satisfy the condition:

$$\lim_{n\to\infty} \mathbf{P}\{\omega : Y_n(\omega) \leqslant a_n\sqrt{n}\} = p, \quad 0 \leqslant p \leqslant 1.$$

353. Let (X_n) be a sequence of independent random variables with equal distributions, variance equal to 1, and $\mathbf{E}[X_n] = 0$ (where $[x]$ represents the integer part of x). Assuming that

$$\lim_{n\to\infty} \mathbf{P}\left\{\omega : \frac{X_1(\omega) + \cdots + X_n(\omega)}{\sqrt{n}} > 0\right\} = \frac{1}{2},$$

calculate $\mathbf{E}\{X_n\} := \mathbf{E}(X_n - [X_n])$.

354. Prove that

$$\lim_{n\to\infty} e^{-n} \sum_{k=0}^{n} \frac{n^k}{k!} = \frac{1}{2}.$$

Hint: Use the Central Limit Theorem for a sequence of independent random variables with Poisson distribution with parameter $\lambda = 1$.

Chapter 8
Extension of Measure

8.1 The Carathéodory Extension Theorem

In this section we describe the Carathéodory Extension Theorem, which states that any pre-measure defined on a given ring $\mathcal{A}$ of subsets of a given set Ω can be extended to a measure on the σ-algebra generated by $\mathcal{A}$, and this extension is unique if the pre-measure is σ-finite. In this statement a "pre-measure" is any finitely additive function $Q\colon \mathcal{A} \to [0, \infty]$ which satisfies the condition

$$\forall\, (A_n)_{n \in \mathbb{N}} \subset \mathcal{A} \quad \bigcup_{n=1}^{\infty} A_n \in \mathcal{A} \implies Q\Big(\bigcup_{n=1}^{\infty} A_n\Big) = \sum_{n=1}^{\infty} Q(A_n).$$

Consequently, any pre-measure on a ring containing all intervals of real numbers can be extended to the Borel algebra of the set of real numbers. This is a very powerful result, and leads, for example, to the proof of the existence of the Lebesgue measure.

This theorem is also known as the *Carathéodory–Fréchet Extension Theorem*, the *Carathéodory–Hopf Extension Theorem*, the *Hopf Extension Theorem* and the *Hahn–Kolmogorov Extension Theorem*.

Let Ω be a space of elementary events. By 2^Ω, we will denote the set of all subsets of the set Ω. We have discussed *measures* many times without introducing their formal definition. Let us now state clearly that a *finite measure* is a set function $\mu\colon \mathcal{F} \to [0, \infty)$ such that $\mu(\Omega) < \infty$ and $\frac{1}{\mu(\Omega)}\mu$ is a probability measure. The Lebesgue measure on a compact set $\Omega \subset \mathbb{R}^k$ is therefore a measure in this sense. Below, we introduce the concept of an even broader class of set functions called *outer measures*.

Definition 8.1 An *outer measure* (*Carathéodory Measure*) on the space Ω is any function $\mu_* : 2^\Omega \to [0, \infty]$ that satisfies the conditions:

(1) $\mu_*(\emptyset) = 0$;
(2) $A \subset B \implies \mu_*(A) \leqslant \mu_*(B)$;
(3) $A_1, A_2, \dots \in \Omega \implies \mu_*\left(\bigcup_{k=1}^\infty A_k\right) \leqslant \sum_{k=1}^\infty \mu_*(A_k)$.

Of course, the last two conditions can be replaced by one:

$$A \subset \bigcup_{k=1}^\infty A_k \subset \Omega \implies \mu_*(A) \leqslant \sum_{k=1}^\infty \mu_*(A_k).$$

Note that the definition of an outer measure differs from the definition of a measure (e.g. probabilistic) only in condition (3). Thus, it can be said that an outer measure is a measure if it is countably additive, i.e., its value on the union of disjoint sets is equal to the sum of its values on these sets.

Examples 8.2 In the following examples, Ω is an arbitrary nonempty set. However, the last example only becomes interesting when Ω contains infinitely many elements.

(a)
$$\mu_*(A) = \begin{cases} 0 & \text{if } A = \emptyset; \\ 1 & \text{if } A \neq \emptyset. \end{cases}$$

(b)
$$\mu_*(A) = \begin{cases} 0 & \text{if } A = \emptyset; \\ \infty & \text{if } A \neq \emptyset. \end{cases}$$

(c)
$$\mu_*(A) = \begin{cases} \text{number of elements of the set } A & \text{if } A \text{ is finite}; \\ \infty & \text{if } A \text{ is infinite}. \end{cases}$$

Definition 8.3 We say that $A \subset \Omega$ is μ_*-*measurable* (satisfies the *Carathéodory condition*) if for every $M \subset \Omega$, the following equality holds:

$$\mu_*(M) = \mu_*(M \cap A) + \mu_*(M \cap A').$$

We will denote the class of μ_*-measurable sets by the symbol $\mathcal{M}$.

8.1 The Carathéodory Extension Theorem

Of course, in the definition of a μ_*-measurable set, it is sufficient to require that for each set $M \subset \Omega$, the following inequality holds:

$$\mu_*(M) \geqslant \mu_*(M \cap A) + \mu_*(M \cap A').$$

The opposite inequality follows from the definition of an outer measure μ_*.

Theorem 8.4 (Carathéodory's Theorem) *Let μ_* be an outer measure on the set Ω and let $\mathcal{M}$ be the class of μ_*-measurable sets. Then, $\mathcal{M}$ is a σ-field of subsets of the set Ω, and the outer measure μ_* restricted to the space $(\Omega, \mathcal{M})$ is a measure, i.e., is a countably additive set function.*

Proof First, we will prove that $\mathcal{M}$ is a σ-field.
Fact 1 If $A \in \mathcal{M}$, it follows from the definition that $A' \in \mathcal{M}$ as well.
Fact 2 $\mathcal{M}$ is a field.
It suffices to show that if $A, B \in \mathcal{M}$, then also $A \cap B \in \mathcal{M}$. Let M be an arbitrary subset of Ω. Then:

$$\mu_*(M) = \mu_*(M \cap A) + \mu_*(M \cap A').$$

Since both sets $M \cap A$ and $M \cap A'$ are subsets of Ω and $B \in \mathcal{M}$,

$$\mu_*(M) = \mu_*(M \cap A \cap B) + \mu_*(M \cap A \cap B')$$
$$+ \mu_*(M \cap A' \cap B) + \mu_*(M \cap A' \cap B').$$

Note that $(A \cap B)' = A \cap B' \cup A' \cap B \cup A' \cap B'$. From Axiom (3) of the outer measure μ_*, we conclude that

$$\mu_*\left(M \cap (A \cap B)'\right) \leqslant \mu_*(M \cap A \cap B') + \mu_*(M \cap A' \cap B) + \mu_*(M \cap A' \cap B').$$

Now, it can be seen that $A \cap B \in \mathcal{M}$ because

$$\mu_*(M) \geqslant \mu_*(M \cap (A \cap B)) + \mu_*(M \cap (A \cap B)').$$

Fact 3 If the sets $A_1, \ldots, A_n \in \mathcal{M}$ are disjoint, then for every $M \subset \Omega$, the following equality holds:

$$\mu_*\left(M \cap \bigcup_{k=1}^{n} A_k\right) = \mu_*(M \cap A_1) + \cdots + \mu_*(M \cap A_n).$$

The proof of this fact is based on the principle of mathematical induction on n. For $n = 2$, we get:

$$\mu_*(M \cap (A_1 \cup A_2)) = \mu_*(M \cap (A_1 \cup A_2) \cap A_1) + \mu_*(M \cap (A_1 \cup A_2) \cap A_1')$$
$$= \mu_*(M \cap A_1) + \mu_*(M \cap A_2)$$

because $(A_1 \cup A_2) \cap A_1 = A_1$ and $(A_1 \cup A_2) \cap A_1' = A_2$. Now, assume that the equality holds true for some $n \in \mathbb{N}$. Then:

$$\mu_*\left(M \cap \bigcup_{k=1}^{n+1} A_k\right)$$

$$= \mu_*\left(\left(M \cap \bigcup_{k=1}^{n+1} A_k\right) \cap A_{n+1}\right) + \mu_*\left(\left(M \cap \bigcup_{k=1}^{n+1} A_k\right) \cap A_{n+1}'\right)$$

$$= \mu_*\left(M \cap A_{n+1}\right) + \mu_*\left(M \cap \bigcup_{k=1}^{n} A_k\right) = \sum_{k=1}^{n+1} \mu_*(M \cap A_k),$$

which follows from the disjointness of the sets $A_1, \ldots, A_{n+1}$ and the inductive assumption.

Fact 4 If $A_1, A_2, \cdots \in \mathcal{M}$ are pairwise disjoint and the set $M \subset \Omega$, then $\mu_*\left(M \cap \bigcup_{k=1}^{\infty} A_k\right) = \sum_{k=1}^{\infty} \mu_*(M \cap A_k)$.

From Axiom (2) of the outer measure and Fact 3, it follows that for every natural number n,

$$\mu_*\left(M \cap \bigcup_{k=1}^{\infty} A_k\right) \geq \mu_*\left(M \cap \bigcup_{k=1}^{n} A_k\right) = \sum_{k=1}^{n} \mu_*(M \cap A_k).$$

Letting n tend to infinity, we get:

$$\mu_*\left(M \cap \bigcup_{k=1}^{\infty} A_k\right) \geq \sum_{k=1}^{\infty} \mu_*(M \cap A_k).$$

The opposite inequality follows from Axiom (3).

Fact 5 $\mathcal{M}$ is a σ-field of sets.

We already know that $\mathcal{M}$ is a field (Fact 2) so it suffices to show that $\mathcal{M}$ is closed under sums of countable families of disjoint sets. Let $A_1, A_2, \cdots \in \mathcal{M}$ be pairwise disjoint. Since $\mathcal{M}$ is a field, $\bigcup_{k=1}^{n} A_k \in \mathcal{M}$ for every $n \in \mathbb{N}$. Note also that for any set $M \subset \Omega$, we have:

$$M \cap \left(\bigcup_{k=1}^{\infty} A_k\right)' \subset M \cap \left(\bigcup_{k=1}^{n} A_k\right)'.$$

Hence,

$$\mu_*(M) = \mu_*\left(M \cap \bigcup_{k=1}^{n} A_k\right) + \mu_*\left(M \cap \left(\bigcup_{k=1}^{n} A_k\right)'\right)$$

8.1 The Carathéodory Extension Theorem

$$\geq \mu_*\left(M \cap \bigcup_{k=1}^{n} A_k\right) + \mu_*\left(M \cap \left(\bigcup_{k=1}^{\infty} A_k\right)'\right)$$

$$= \sum_{k=1}^{n} \mu_*(M \cap A_k) + \mu_*\left(M \cap \left(\bigcup_{k=1}^{\infty} A_k\right)'\right).$$

Letting n tend to infinity, we get:

$$\mu_*(M) \geq \sum_{k=1}^{\infty} \mu_*(M \cap A_k) + \mu_*\left(M \cap \left(\bigcup_{k=1}^{\infty} A_k\right)'\right)$$

$$= \mu_*\left(M \cap \bigcup_{k=1}^{\infty} A_k\right) + \mu_*\left(M \cap \left(\bigcup_{k=1}^{\infty} A_k\right)'\right),$$

which implies that $\bigcup_{k=1}^{\infty} A_k \in \mathcal{M}$.

Fact 6 The outer measure μ_* as a function on $(\Omega, \mathcal{M})$ is a measure.

Assume that the sets $A_1, A_2, \cdots \in \mathcal{M}$ are pairwise disjoint. From Fact 4, it follows that for any $M \subset \Omega$,

$$\mu_*\left(M \cap \bigcup_{k=1}^{\infty} A_k\right) = \sum_{k=1}^{\infty} \mu_*(M \cap A_k).$$

To show the countable additivity of an external measure μ_* on $\mathcal{M}$, it suffices to take $M = \Omega$. This ends the proof of Carathéodory's Theorem. □

It is worth noting here that the outer measure μ_* restricted to $(\Omega, \mathcal{M})$ is a complete measure, i.e., such that every subset of any null set is a measurable null set. In particular, the σ-field $\mathcal{M}$ contains all subsets of null sets. Indeed, if $A \subset B \in \mathcal{M}$ and $\mu_*(B) = 0$, then by Axiom (2) $0 \leq \mu_*(A) \leq \mu_*(B) = 0$. At the same time, if $\mu_*(A) = 0$, then for any $M \subset \Omega$, $\mu_*(M \cap A) \leq \mu_*(A) = 0$. Hence, $A \in \mathcal{M}$ because

$$\mu_*(M) \geq \mu_*(M \cap A') = \mu_*(M \cap A) + \mu_*(M \cap A').$$

Carathéodory's Theorem is a very useful tool for probability calculus and stochastic processes. It is usually difficult to define a measure by giving its value on every set of a fixed σ-field, while it is relatively easier to define its values on the sets of a class which generates the given σ-field.

The following theorem, based on Carathéodory's Theorem, concludes that a countably additive set function on a field $\mathcal{A}$ extends uniquely to a measure on the σ-field generated by $\mathcal{A}$. It is easy to see that the assumptions of this theorem can be further weakened; for example, we can assume that the countably additive set function is specified on a ring $\mathcal{A}$ of subsets of Ω if Ω is the countable union of the elements of $\mathcal{A}$.

Theorem 8.5 (Measure Extension Theorem) *Assume that α is a pre-measure, i.e., it is a finite, countably additive, non-negative set function defined on a field $\mathcal{A}$ of subsets of the set Ω, i.e., if $A_1, A_2, \cdots \in \mathcal{A}$ are disjoint and such that $\bigcup_{n=1}^{\infty} A_n \in \mathcal{A}$, then*

$$\alpha\left(\bigcup_{n=1}^{\infty} A_n\right) = \sum_{n=1}^{\infty} \alpha(A_n).$$

Then, there exists exactly one measure μ on $\sigma(\mathcal{A})$ such that $\alpha(A) = \mu(A)$ for every set $A \in \mathcal{A}$.

Proof The main difficulty of the proof is to construct an appropriate outer measure and use Carathéodory's Theorem. We define this outer measure as follows:

$$\mu_*(A) \stackrel{\text{def}}{=} \inf\left\{\sum_{k=1}^{\infty} \alpha(A_k) : A \subset \bigcup_{k=1}^{\infty} A_k, \ A_1, A_2, \cdots \in \mathcal{A}\right\}$$

for every $A \subset \Omega$. Let us prove that μ_* is an outer measure on Ω. Since $\alpha(A_k) \geqslant 0$ for $k = 1, 2, \ldots$, we have $\mu_*(A) \geqslant 0$, and since $\emptyset \subset \emptyset \cup \emptyset \cup \ldots$,

$$0 \leqslant \mu_*(\emptyset) \leqslant \alpha(\emptyset) + \alpha(\emptyset) + \cdots = 0,$$

which means that $\mu_*(\emptyset) = 0$. It remains to show that the following implication holds:

$$E \subset \bigcup_{k=1}^{\infty} E_k \implies \mu_*(E) \leqslant \sum_{k=1}^{\infty} \mu_*(E_k).$$

Note that if $A \subset \Omega$, then $A \subset \Omega \cup \emptyset \cup \ldots$, thus $\mu_*(A) \leqslant \alpha(\Omega) < \infty$. By definition of μ_*, for every $k \in \mathbb{N}$, there exists a sequence of sets $A_{k,1}, A_{k,2}, \cdots \in \mathcal{A}$ such that

$$E_k \subset \bigcup_{i=1}^{\infty} A_{k,i}, \quad \mu_*(E_k) + \frac{\varepsilon}{2^k} \geqslant \sum_{i=1}^{\infty} \alpha(A_{k,i}).$$

As $E \subset \bigcup_{k=1}^{\infty} E_k \subset \bigcup_{k=1}^{\infty} \bigcup_{i=1}^{\infty} A_{k,i}$,

$$\mu_*(E) \leqslant \sum_{k=1}^{\infty} \sum_{i=1}^{\infty} \alpha(A_{k,i}) \leqslant \sum_{k=1}^{\infty} \mu_*(E_k) + \varepsilon,$$

which, given that ε is arbitrary, implies that $\mu_*(E) \leqslant \sum_{k=1}^{\infty} \mu_*(E_k)$.

8.1 The Carathéodory Extension Theorem

Now, we can use Carathéodory's Theorem for the external measure μ_* and conclude that μ_* restricted to the σ-field $\mathcal{M}$ of μ_*-measurable sets is a countably additive measure. Let us prove that $\mathcal{A} \subset \mathcal{M}$ and that μ_* coincides with α on the field $\mathcal{A}$:

Let $A \in \mathcal{A}$, $M \subset \Omega$ and $A_1, A_2, \cdots \in \mathcal{A}$ be a sequence of sets such that $M \subset \bigcup_{k=1}^{\infty} A_k$. Of course, $A_k \cap A \in \mathcal{A}$, $A_k \cap A' \in \mathcal{A}$ and $M \cap A \subset \bigcup_{k=1}^{\infty}(A_k \cap A)$, $M \cap A' \subset \bigcup_{k=1}^{\infty}(A_k \cap A')$. From the definition of the function μ_*, we get

$$\mu_*(M \cap A) + \mu_*(M \cap A') \leqslant \sum_{k=1}^{\infty} \alpha(A_k \cap A) + \sum_{k=1}^{\infty} \alpha(A_k \cap A')$$

$$= \sum_{k=1}^{\infty} [\alpha(A_k \cap A) + \alpha(A_k \cap A')] = \sum_{k=1}^{\infty} \alpha(A_k).$$

Since the sequence $A_1, A_2, \ldots$ is arbitrarily chosen, we conclude that

$$\mu_*(M \cap A) + \mu_*(M \cap A') \leqslant \inf \left\{ \sum_{k=1}^{\infty} \alpha(A_k) : M \subset \bigcup_{k=1}^{\infty} A_k \right\} = \mu_*(M),$$

thus $A \in \mathcal{M}$, consequently $\sigma(\mathcal{A}) \subset \mathcal{M}$, which was to be shown.

Let us take any set $A \in \mathcal{A}$. Since $A \subset A \cup \emptyset \cup \emptyset \cup \ldots$, from the definition of the measure μ_*, we get

$$\mu_*(A) \leqslant \alpha(A) + \alpha(\emptyset) + \alpha(\emptyset) + \cdots = \alpha(A).$$

To prove the opposite inequality, let us consider any sequence of sets $A_k \in \mathcal{A}$, $k = 1, 2, \ldots$ such that $A \subset \bigcup_{k=1}^{\infty} A_k$. The sets $B_1 = A \cap A_1$, $B_{n+1} = A \cap [A_{n+1} \setminus \bigcup_{k=1}^{n} A_k]$, $n \in \mathbb{N}$ form a sequence of pairwise disjoint sets such that $A = \bigcup_{k=1}^{\infty} B_k$ and $B_k \in \mathcal{A}$ for $k \in \mathbb{N}$. Thus:

$$\alpha(A) = \sum_{k=1}^{\infty} \alpha(B_k) \leqslant \sum_{k=1}^{\infty} \alpha(A_k),$$

which, given that the sequence $A_1, A_2, \ldots$ is freely chosen, implies the opposite inequality: $\alpha(A) \leqslant \mu_*(A)$.

Let μ denote the restriction of the outer measure μ_* to the σ-field $\sigma(\mathcal{A})$. To complete the proof, it is still necessary to prove that μ is the only measure that satisfies the conditions of the theorem. Suppose this is not the case. Then there exists another measure λ on $(\Omega, \sigma(\mathcal{A}))$ such that $\alpha(A) = \lambda(A)$ for each set $A \in \mathcal{A}$. Let

$$\mathcal{G} = \left\{ E \in \sigma(\mathcal{A}) : \lambda(E) = \alpha(E) \right\}.$$

By assumption, $\mathcal{A} \subset \mathcal{G}$. If $E \in \mathcal{G}$, then $\lambda(E') = \lambda(\Omega) - \lambda(E) = \alpha(\Omega) - \alpha(E) = \alpha(E')$, therefore $E' \in \mathcal{G}$. If the sets $E_1, E_2, \cdots \in \mathcal{G}$ are pairwise disjoint, then

$$\lambda\left(\bigcup_{k=1}^{\infty} E_k\right) = \sum_{k=1}^{\infty} \lambda(E_k) = \sum_{k=1}^{\infty} \alpha(E_k) = \alpha\left(\bigcup_{k=1}^{\infty} E_k\right),$$

which shows that $\bigcup_{k=1}^{\infty} E_k \in \mathcal{G}$. Hence, it follows that $\mathcal{G}$ is a σ-field containing $\mathcal{A}$ and contained in $\sigma(\mathcal{A})$. Consequently, $\mathcal{G} = \sigma(\mathcal{A})$ and $\lambda = \mu$, which contradicts our assumptions. □

8.1.1 Exercises

355. Prove that the sum of countably many outer measures is also an outer measure.
356. Let μ be a probability measure on $(\Omega, \mathcal{F})$ and A_k, B_k for $k \in \mathbb{N}$ be subsets of Ω. Prove that if the outer measure μ_* has the property $\mu_*(A_k \triangle B_k) = 0$ for every $k \in \mathbb{N}$, then

$$\mu_*\left(\bigcup_{k=1}^{\infty} A_k\right) = \mu_*\left(\bigcup_{k=1}^{\infty} B_k\right).$$

357. Let μ be a probability measure on the space $(\Omega, \mathcal{F})$ and let $E \subset \Omega$,

$$\mu^*(E) \stackrel{\text{def}}{=} \inf\{\mu(A) \colon E \subset A, A \in \mathcal{F}\}.$$

Prove that μ^* is an outer measure on 2^{Ω} and that $\mu^*\big|_{\mathcal{F}} = \mu$.

358. Let μ be a probability measure on the space $(\Omega, \mathcal{F})$ and let μ^* be the outer measure defined in the previous exercise. Let $\overline{\mathcal{F}}$ denote the σ-field of sets which are μ^*-measurable, and let $\overline{\mu}$ denote the restriction of μ^* to the σ-field $\overline{\mathcal{F}}$. Prove that $\overline{\mu}$ is a complete measure which coincides with μ on $\mathcal{F}$.

359. Let α be a finite non-negative additive set function on the field $\mathcal{A} \subset \Omega$. For $E \subset \Omega$, we define:

$$\alpha_*(E) := \inf\left\{\sum_{k=1}^{n} \alpha(A_k) \colon A_1, \ldots, A_n \in \mathcal{A}, E \subset \bigcup_{k=1}^{n} A_k\right\}.$$

Prove that α^* is an outer measure.

360. Let $\Omega = \mathbb{N}$. For any set $A \subset \Omega$, we define $\#(A, n)$ as the number of elements of A that are less than n. Let C be the class of sets $A \subset \Omega$ for which the following limit exists

$$d(A) \stackrel{\text{def}}{=} \lim_{n \to \infty} \frac{\#(A, n)}{n}.$$

The function $d(A)$ is called the *density* of the set A.

(a) Show that $d : C \to [0, 1]$ is an additive but not countably additive set function.
(b) Prove that C is not a field of sets.

Hint. Let A be the set of even numbers. Let B be the set of even numbers between 2^{2n} and 2^{2n+1} and odd numbers between 2^{2n-1} and 2^{2n} for every $n \in \mathbb{N}$. Which of the sets A, B and $A \cap B$ belong to C?

8.2 Cumulative Distribution Functions

Let us return to Theorem 3.17, which states that a function F that satisfies the three given conditions is the distribution function of some random variable X. It also means that the function F is the distribution function of the probability measure $\mathbf{P}_X$, which is the distribution of the variable X. The proof of Theorem 3.17 presented earlier was based on the construction of a suitable random variable. The proof of the theorem presented below is based on the construction of an appropriate probability measure on the space $(\mathbb{R}, \mathcal{B})$.

Theorem 8.6 *Suppose the function* $F : \mathbb{R} \to [0, 1]$ *satisfies the following conditions:*

(1) *F is a nondecreasing function;*
(2) $\lim_{t \to -\infty} F(t) = 0$, $\lim_{t \to \infty} F(t) = 1$;
(3) *F is left-continuous, i.e., $F(t) = \lim_{s \nearrow t} F(s)$ for every $t \in \mathbb{R}$.*

Then, there exists exactly one probability measure $\mathbf{P}$ on $(\mathbb{R}, \mathcal{B}(\mathbb{R}))$ such that for $t \in \mathbb{R}$

$$F(t) = \mathbf{P}((-\infty, t)).$$

Proof If such a probability measure $\mathbf{P}$ existed, then for any $s < t$ we would have $F(t) = \mathbf{P}((-\infty, s)) + \mathbf{P}([s, t)) = F(s) + \mathbf{P}([s, t))$. Following this remark, we define the $\mathbf{P}$ measure on the class of sets

$$\mathcal{F} := \big\{ [s, t) : s, t \in \mathbb{R}, s < t \big\}$$

by the formula

$$\mathbf{P}([s, t)) := F(t) - F(s).$$

□

Lemma 8.7 *If the sets $A_1, A_2, \ldots, A_n \in \mathcal{F}$ are disjoint and $\bigcup_{i=1}^n A_i \subset A_0$ for some $A_0 = [a_0, b_0) \in \mathcal{F}$, then*

$$\mathbf{P}(A_0) \geqslant \sum_{i=1}^n \mathbf{P}(A_i).$$

Proof of Lemma 8.7 Since the intervals $A_i = [a_i, b_i)$ are disjoint and A_0 contains their union, we can renumber them in such a way that

$$a_0 \leqslant a_1 \leqslant b_1 \leqslant a_2 \leqslant \cdots \leqslant a_n \leqslant b_n \leqslant b_0.$$

By property (1), the function F is nondecreasing; thus

$$\mathbf{P}(A_0) = [F(b_0) - F(b_n)] + [F(b_n) - F(a_n)] + [F(a_n) - F(b_{n-1})]$$
$$+ \cdots + [F(a_2) - F(b_1)] + [F(b_1) - F(a_1)] + [F(a_1) - F(a_0)]$$
$$= [F(b_0) - F(b_n)] + \sum_{i=1}^n [F(b_i) - F(a_i)] + \sum_{i=2}^{n-1} [F(a_i) - F(b_{i-1})]$$
$$+ [F(a_1) - F(a_0)] \geqslant \sum_{i=1}^n [F(b_i) - F(a_i)] = \sum_{i=1}^n \mathbf{P}(A_i).$$

□

Lemma 8.8 *$\mathbf{P}$ is a countably additive set function on $\mathcal{F}$.*

Proof of Lemma 8.8 Assume that $A_0 = [a_0, b_0) = \bigcup_{i=1}^\infty A_i$ for pairwise disjoint sets $A_i = [a_i, b_i) \in \mathcal{F}$. Since for every $n \in \mathbb{N}$ $\bigcup_{i=1}^n A_i \subset A_0$, by Lemma 8.7, we have:

$$\mathbf{P}(A_0) \geqslant \sum_{i=1}^n \mathbf{P}(A_i).$$

Hence, it already follows that

$$\mathbf{P}(A_0) \geqslant \lim_{n \to \infty} \sum_{i=1}^n \mathbf{P}(A_i) = \sum_{i=1}^\infty \mathbf{P}(A_i). \qquad (*)$$

8.2 Cumulative Distribution Functions

To prove that the opposite inequality is also true, let $\varepsilon \in (0, b_0 - a_0)$. The left-continuity of the function F shows that for every $i \in \mathbb{N}$, one can find a real number $a_i' < a_i$ for which $F(a_i) - F(a_i') < \varepsilon/2^i$. The compact interval $[a_0, b_0 - \varepsilon]$ is a subset of A_0, thus

$$[a_0, b_0 - \varepsilon] \subset \bigcup_{i=1}^{\infty} [a_i, b_i) \subset \bigcup_{i=1}^{\infty} (a_i', b_i).$$

The Heine–Borel Theorem applied to a subset S of $\mathbb{R}^n$ asserts that the following two statements are equivalent:

- S is closed and bounded,
- S is compact, that is, any covering of S by a collection of open sets contains a finite subcovering.

By this theorem, there exists an $n \in \mathbb{N}$ such that

$$[a_0, b_0 - \varepsilon] \subset \bigcup_{i=1}^{n} (a_i', b_i).$$

We renumber the set $\{(a_i', b_i) : i = 1, 2, \ldots, n\}$ in such a way that $a_0 \in (a_1', b_1)$, $b_1 \in (a_2', b_2)$, $b_2 \in (a_3', b_3)$, etc. We will finally find a number $k \leqslant n$ such that $b_0 - \varepsilon \in (a_k', b_k)$. If not all the elements $\{(a_i', b_i) : i = 1, 2, \ldots, n\}$ are used in this construction, the remaining elements are numbered $k+1, \ldots, n$. Hence, we get:

$$F(b_0) - F(a_0) = F(b_0) - F(b_0 - \varepsilon) + F(b_0 - \varepsilon) - F(a_0)$$

$$\leqslant F(b_0) - F(b_0 - \varepsilon) + F(b_k) - F(a_1')$$

$$\leqslant F(b_0) - F(b_0 - \varepsilon) + \sum_{i=1}^{n} \left(F(b_i) - F(a_i') \right)$$

$$\leqslant F(b_0) - F(b_0 - \varepsilon) + \sum_{i=1}^{n} \left(F(b_i) - F(a_i) \right) + \sum_{i=1}^{n} \frac{\varepsilon}{2^i}$$

$$\leqslant F(b_0) - F(b_0 - \varepsilon) + \sum_{i=1}^{\infty} \left(F(b_i) - F(a_i) \right) + \varepsilon.$$

All we need to do now is to apply left-continuity of the function F, i.e., the condition $\lim_{\varepsilon \to 0} F(b_0 - \varepsilon) = F(b_0)$, to obtain that the inequality opposite to (∗) holds. Consequently,

$$\mathbf{P}(A_0) = \sum_{i=1}^{\infty} \mathbf{P}(A_i).$$

□

Now, let $\mathcal{F}_0$ be the class of sets defined as follows:

$$\mathcal{F}_0 := \{A_1 \cup \cdots \cup A_n : n \in \mathbb{N}, A_1, \ldots, A_n \in \mathcal{F}, A_i \cap A_j = \emptyset \text{ for } i \neq j\}.$$

Lemma 8.9 $\mathcal{F}_0$ *is a ring of sets.*

Proof of Lemma 8.9 We need to show that if $A, B \in \mathcal{F}_0$, then the union $A \cup B$ and difference $A \setminus B$ also belong to $\mathcal{F}_0$. Let then $A = E_1 \cup \cdots \cup E_n \in \mathcal{F}_0$ and $B = G_1 \cup \cdots \cup G_m \in \mathcal{F}_0$, where $E_i, G_j \in \mathcal{F}$ and $E_i \cap E_j = G_i \cap G_j = \emptyset$ for $i \neq j$.

If $A \cap B = \emptyset$, then also $E_i \cap G_j = \emptyset$ for $i \neq j$. This implies that $A \cup B$ is a finite union of disjoint sets from the class $\mathcal{F}$, thus it belongs to $\mathcal{F}_0$.

Now consider the difference $A \setminus B$ without assuming that the sets are disjoint. We obtain

$$A \setminus B = \bigcup_{i=1}^{n} E_i \setminus \bigcup_{j=1}^{m} G_j = \bigcup_{i=1}^{n} \left(E_i \setminus \bigcup_{j=1}^{m} G_j \right).$$

To complete the proof, it must be shown by mathematical induction that for every $E = [a, b)$ and $G_1, \ldots, G_m \in \mathcal{F}$, the difference $E \setminus \bigcup_{j=1}^{n} G_j$ belongs to $\mathcal{F}_0$. For $n = 1$ and $G_1 = [c, d)$, we only need to consider all possible cases of the relative positions of the numbers a, b, c and d to state that $E \setminus G_1$ can be represented as the union of at most two (possibly empty) sets from $\mathcal{F}$. Hence, it is easy to get that $E \setminus \bigcup_{j=1}^{m} G_j$ is the union of at most 2^n of the sets from $\mathcal{F}$. □

We now extend the definition of the set function **P** to the class $\mathcal{F}_0$:

$$\left(A = A_1 \cup \cdots \cup A_n, A_i \in \mathcal{F}, A_i \cap A_j = \emptyset \text{ if } i \neq j \right) \implies \mathbf{P}(A) \stackrel{\text{def}}{=} \sum_{i=1}^{n} \mathbf{P}(A_i).$$

Lemma 8.10 *The extension of* **P** *to* $\mathcal{F}_0$ *is well defined.*

Proof of Lemma 8.10 Let us consider two different representations of the set $A \in \mathcal{F}_0$ as the union of disjoint sets from $\mathcal{F}$:

$$A = \bigcup_{i=1}^{n} A_i = \bigcup_{j=1}^{m} B_j, \quad A_i, B_j \in \mathcal{F}, A_i \cap A_k = B_i \cap B_k = \emptyset \text{ for } i \neq k.$$

It is easy to verify that the class $\mathcal{F}$ contains all the intersections of its elements, so for any $i = 1, \ldots, n, j = 1, \ldots, m$, we have $A_i \cap A_j \in \mathcal{F}$. Since the set function **P** is countably additive on $\mathcal{F}$, we have

$$\sum_{i=1}^{n} \mathbf{P}(A_i) = \sum_{i=1}^{n} \mathbf{P}\left(A_i \cap \bigcup_{j=1}^{m} B_j \right)$$

8.2 Cumulative Distribution Functions

$$= \sum_{i=1}^{n} \mathbf{P}\left(\bigcup_{j=1}^{m}(A_i \cap B_j)\right)$$

$$= \sum_{i=1}^{n}\sum_{j=1}^{m} \mathbf{P}(A_i \cap B_j).$$

We also have:

$$\sum_{j=1}^{m} \mathbf{P}(B_j) = \sum_{j=1}^{m} \mathbf{P}\left(B_j \cap \bigcup_{i=1}^{n} A_i\right)$$

$$= \sum_{j=1}^{m} \mathbf{P}\left(\bigcup_{i=1}^{n}(A_i \cap B_j)\right)$$

$$= \sum_{j=1}^{m}\sum_{i=1}^{n} \mathbf{P}(A_i \cap B_j).$$

The two sums differ only in the order of the components, so we have the equality $\sum_{i=1}^{n} \mathbf{P}(A_i) = \sum_{j=1}^{m} \mathbf{P}(B_j)$ and the extension of the set function $\mathbf{P}$ to $\mathcal{F}_0$ is well defined. □

Lemma 8.11 $\mathbf{P}$ *is a σ-additive set function on $\mathcal{F}_0$.*

Proof of Lemma 8.11 Let $A_1, A_2, \ldots$ be disjoint sets belonging to the ring $\mathcal{F}_0$ such that their union $A_0 = \bigcup_{i=1}^{\infty} A_i$ also belongs to $\mathcal{F}_0$. Each of these sets can be written as a finite union of disjoint sets belonging to the class $\mathcal{F}$, i.e.,

$$\forall i = 0, 1, \ldots \quad A_i = \bigcup_{j=1}^{m_i} E_{i,j}, \quad E_{i,j} \in \mathcal{F}.$$

Since the class $\mathcal{F}$ is closed under intersections and the set function $\mathbf{P}$ is σ-additive on $\mathcal{F}$, we have

$$\mathbf{P}(A_0) = \sum_{j=1}^{m_0} \mathbf{P}(E_{0,j}) = \sum_{j=1}^{m_0} \mathbf{P}\left(E_{0,j} \cap \bigcup_{i=1}^{\infty} A_i\right)$$

$$= \sum_{j=1}^{m_0} \mathbf{P}\left(\bigcup_{i=1}^{\infty}(E_{0,j} \cap A_i)\right) = \sum_{j=1}^{m_0} \mathbf{P}\left(\bigcup_{i=1}^{\infty}\bigcup_{k=1}^{m_i}(E_{0,j} \cap E_{i,k})\right)$$

$$= \sum_{j=1}^{m_0}\sum_{i=1}^{\infty}\sum_{k=1}^{m_i} \mathbf{P}(E_{0,j} \cap E_{i,k}) = \sum_{i=1}^{\infty}\sum_{k=1}^{m_i}\sum_{j=1}^{m_0} \mathbf{P}(E_{0,j} \cap E_{i,k})$$

$$= \sum_{i=1}^{\infty}\sum_{k=1}^{m_i} \mathbf{P}(E_{i,k}) = \sum_{i=1}^{\infty} \mathbf{P}(A_i).$$

□

Going back to the proof of Theorem 8.6, we use the measure extension Theorem 8.5 formulated for a ring of sets. Since $\mathcal{F}_0$ is a ring of sets and $\mathbb{R}$ can be represented as the countable union of sets from $\mathcal{F}_0$, every countably additive function **P** on $\mathcal{F}_0$ can be uniquely extended to a countably additive set function **P**′ on the σ-field $\sigma(\mathcal{F}_0)$ generated by $\mathcal{F}_0$.

We have yet to show that $\sigma(\mathcal{F}_0)$ is a σ-field containing Borel sets on the straight line. It is enough to check that the family $\sigma(\mathcal{F}_0)$ contains open intervals and is closed under complements. Note that for any $a, b \in \mathbb{R}, a < b$,

$$(a, b) = \bigcup_{n=1}^{\infty} \left[a - \frac{b-a}{2^n}, b\right) \implies (a, b) \in \sigma(\mathcal{F}_0).$$

In a similar way, we obtain

$$\mathbb{R} = \bigcup_{k=-\infty}^{\infty} [k, k+1) \implies \mathbb{R} \in \sigma(\mathcal{F}_0).$$

Since the σ-ring $\sigma(\mathcal{F}_0)$ is closed under taking the difference of two sets, and $\mathbb{R} \in \sigma(\mathcal{F}_0)$, $\sigma(\mathcal{F}_0)$ is also closed under taking the complement operation. Therefore, $\sigma(\mathcal{F}_0)$ is a σ-field.

□

8.3 The Radon–Nikodym Theorem

In this section, we will discuss a theorem which is important for the theory of measure, probability theory and stochastic processes, the Radon–Nikodym Theorem. Due to the later applications of this theorem, it is necessary to formulate it in a rather general form, i.e., for measures that do not have to be finite.

Consider a probability space $(\Omega, \mathcal{F}, \mu)$. The measure μ is a σ-*finite measure* on $\mathcal{F}$ if the following conditions are satisfied:

(1) $\mu(A) \geq 0$ for every $A \in \mathcal{F}$;
(2) if $A_i \in \mathcal{F}$ for $i \in \mathbb{N}$ are pairwise disjoint, then $\mu\left(\bigcup_{i=1}^{\infty} A_i\right) = \sum_{i=1}^{\infty} \mu(A_i)$;
(3) there exists a sequence of sets $A_i \in \mathcal{F}, i \in \mathbb{N}$, such that $\Omega = \bigcup_{i=1}^{\infty} A_i$ and $\mu(A_i) < \infty$ for every $i \in \mathbb{N}$.

8.3 The Radon–Nikodym Theorem

An example of a σ-finite measure is the well-known Lebesgue measure on $(\mathbb{R}, \mathcal{B})$. For the sets A_n, we can take intervals $[-n, n]$.

Assume that $f : \Omega \to \mathbb{R}$ is a measurable function, μ is a σ-finite measure on $(\Omega, \mathcal{F})$ and the sequence of sets (A_n) is such that $\Omega = \bigcup_{i=1}^{\infty} A_i$, $\mu(A_n) < \infty$ for all $n \in \mathbb{N}$. We know the definition of the integral $\int_\Omega f d\mu$ for a probability measure μ. For a σ-finite measure, we shall define it as follows:

$$\int_\Omega f(\omega)\mu(d\omega) \stackrel{\text{def}}{=} \lim_{n\to\infty} \int_{B_n} f(\omega)\mu(d\omega) = \lim_{n\to\infty} \mu(B_n) \cdot \int_{B_n} f(\omega) \frac{\mu(d\omega)}{\mu(B_n)},$$

where $B_n = \bigcup_{i=1}^n A_i$. This definition applies only if the right side of this formula exists. The integrals over the sets B_n are well defined, because $\mu(B_n) < \infty$, thus we are integrating with respect to a probability measure $\frac{\mu(d\omega)}{\mu(B_n)}$. If the integral $\int_\Omega |f(\omega)|\mu(d\omega)$ exists, i.e., the limit of the integrals $\int_{B_n} f(\omega)\mu(d\omega)$ exists, we say that the function f is *integrable with respect to the σ-finite measure μ*.

Lemma 8.12 *Let $f : \Omega \to \mathbb{R}$ be a function which is integrable with respect to a σ-finite measure μ on the space $(\Omega, \mathcal{F})$. If for every $A \in \mathcal{F}$*

$$\int_A f(\omega)\mu(d\omega) \geq 0,$$

then $f \geq 0$ μ-almost everywhere.

Proof Let $A = \{\omega : f(\omega) < 0\}$ and, let $A_n = \{\omega : f(\omega) < -\frac{1}{n}\}$, $n \in \mathbb{N}$. The sequence (A_n) is increasing and $A = \bigcup_{n=1}^{\infty} A_n$. By our assumptions, we have

$$0 \leq \int_{A_n} f(\omega)\mu(d\omega) \leq -\frac{1}{n}\mu(A_n),$$

which implies that $\mu(A_n) = 0$ for every $n \in \mathbb{N}$. Hence, the result follows because

$$\mu(A) = \lim_{n\to\infty} \mu(A_n) = 0.$$

□

Sometimes we will need to consider a signed measure, i.e., the difference of two measures with separate supports. The following definition describes this more precisely:

Definition 8.13 We say that μ is a *σ-finite signed measure* on $(\Omega, \mathcal{F})$ if there exist a set $B \in \mathcal{F}$ and two σ-finite measures μ^+, μ^- on $(\Omega, \mathcal{F})$ such that one of them is a finite measure, and

$$\mu^+(B') = 0, \quad \mu^-(B) = 0, \quad \forall A \in \mathcal{B} \quad \mu(A) = \mu^+(A) - \mu^-(A).$$

If both measures μ_+ and μ_- are finite, we will say that μ is a *signed measure*.

The requirement that at least one of the measures μ^+ or μ^- should be finite guarantees that the value of the measure μ on every set from the σ-field $\mathcal{F}$ is well defined because any indeterminate term of the type $\infty - \infty$ cannot appear. Hence, the set function

$$\kappa(A) = \int_A \frac{x}{1+x^2}\, dx, \quad A \in \mathcal{B}$$

is not a signed measure, even though its restriction to any bounded subset $A \in \mathcal{B}$ is a finite signed measure.

If the integrals $\int |f(\omega)|\mu_+(d\omega)$ and $\int |f(\omega)|\mu_-(d\omega)$ exist and at least one of them is finite, we say that the function f is *integrable with respect to the σ-finite signed measure* μ.

Definition 8.14 Let ν be a signed measure and let μ be a non-negative σ-finite measure on $(\Omega, \mathcal{F})$. The measure ν is *absolutely continuous* with respect to the measure μ (notation: $\nu \ll \mu$) if for any set $A \in \mathcal{F}$, the following implication holds:

$$\mu(A) = 0 \implies \nu(A) = 0.$$

Example 8.15 Let X be a continuous type random variable on $(\Omega, \mathcal{F}, \mathbf{P})$ and let f be the density function for X. Then,

$$\mathbf{P}_X(A) = \int_A f(x)\, dx \quad A \in \mathcal{B}.$$

If $\lambda_1(A) = 0$, where λ_1 is the Lebesgue measure on $\mathbb{R}$, then $\mathbf{P}_X(A) = 0$, hence $\mathbf{P}_X$ is absolutely continuous with respect to λ_1. This is where the term *continuous type variable* comes from.

The next, important theorem is presented here without proof.

Theorem 8.16 (Radon–Nikodym Theorem) *Let ν be a signed measure and let μ be a non-negative σ-finite measure on $(\Omega, \mathcal{F})$. If $\nu \ll \mu$, then there exists an $\mathcal{F}$-measurable, μ-integrable function $f : \Omega \to \mathbb{R}$ such that for any $A \in \mathcal{F}$, we have*

$$\nu(A) = \int_A f(\omega)\mu(d\omega).$$

The function f is uniquely determined up to a set of μ-measure zero. We call the function f the Radon–Nikodym derivative *of the measure ν with respect to the measure μ, and we denote it by $f = \frac{d\nu}{d\mu}$.*

Remark 8.17 Note that the uniqueness of the function f follows from Lemma 8.12. Indeed, if there were two such functions $f_1 \neq f_2$, then for any $A \in \mathcal{F}$ we would have

$$\int_A f_1(\omega)\mu(d\omega) = \int_A f_2(\omega)\mu(d\omega).$$

Now all we need to do is to apply Lemma 8.12 to the function $f = \pm(f_1 - f_2)$ to get $\mu\{\omega : f_1(\omega) \neq f_2(\omega)\} = 0$.

8.3.1 Exercises

361. Let X be a random variable on $(\Omega, \mathcal{F}, \mathbf{P})$ and let λ_1 be the Lebesgue measure on $\mathbb{R}$. Prove that the distribution $\mathbf{P}_X$ is of continuous type iff $\mathbf{P}_X \ll \lambda$.
362. Find the Radon–Nikodym derivative of the distribution $\Gamma(p, b)$ with respect to: (a) the normal distribution $N(0, 1)$; (b) the exponential distribution $\Gamma(1, \lambda)$.
363. Find the Radon–Nikodym derivative of the Poisson distribution with parameter λ with respect to: (a) the geometric distribution with parameter $p \in (0, 1)$; (b) the Poisson distribution with parameter $\alpha > \lambda$.
364. Assume that probability measures μ and ν are absolutely continuous with respect to the Lebesgue measure. Does this mean that either $\mu \ll \nu$ or $\nu \ll \mu$ holds?
365. Assume that a discrete measure μ has atoms $\{x_1, x_2, \ldots\}$, and a measure ν is absolutely continuous with respect to μ. What can be said about the support of the measure ν?
366. Is the normal distribution absolutely continuous with respect to some exponential distribution?

8.4 Conditional Expectation

Definition 8.18 Let X be an integrable random variable on a probability space $(\Omega, \mathcal{F}, \mathbf{P})$ and let $\mathcal{A}$ be a sub-σ-field of the σ-field $\mathcal{F}$. A random variable $\mathbf{E}(X|\mathcal{A})$ is a *conditional expectation* of the variable X given the σ-field $\mathcal{A}$ if the following conditions are satisfied:

(i) $\mathbf{E}(X|\mathcal{A})$ is an $\mathcal{A}$-measurable function;
(ii) for every $A \in \mathcal{A}$

$$\int_A X \, d\mathbf{P} = \int_A \mathbf{E}(X|\mathcal{A}) d\mathbf{P}.$$

Theorem 8.19 *For every integrable random variable X on $(\Omega, \mathcal{F}, \mathbf{P})$ and any sub-σ-field $\mathcal{A}$ of the σ-field $\mathcal{F}$, there exists a conditional expectation $\mathbf{E}(X|\mathcal{A})$. Moreover, the variable $\mathbf{E}(X|\mathcal{A})$ is uniquely determined up to a set of measure zero.*

Proof Since the random variable X is integrable, the following function

$$\nu(A) = \int_A X \, d\mathbf{P}, \quad A \in \mathcal{A}$$

is a finite signed measure on the space $(\Omega, \mathcal{A})$. If $\mathbf{P}(A) = 0$, then also $\nu(A) = 0$, hence, $\nu \ll \mathbf{P}$. By the Radon–Nikodym Theorem, there exists an $\mathcal{A}$-measurable function $f = \frac{d\nu}{d\mathbf{P}}$, such that

$$\nu(A) = \int_A f \, d\mathbf{P} \quad A \in \mathcal{A}.$$

Hence, it is enough to define $\mathbf{E}(X|\mathcal{A}) := f$. □

8.4.1 Conditional Expectation Properties

Below we will discuss the basic properties of conditional expectation. We assume that the random variables discussed here have a finite expected value. The first six properties are simple corollaries of the definition and the properties of integral, therefore their proofs are omitted.

Property 1 *If $\mathcal{A} = \{\emptyset, \Omega\}$, then $\mathbf{E}(X|\mathcal{A}) = \mathbf{E}X$ a.e.*

Property 2 *If X is an $\mathcal{A}$-measurable variable, then $\mathbf{E}(X|\mathcal{A}) = X$ a.e.*

Property 3 *If $X \geqslant 0$, then $\mathbf{E}(X|\mathcal{A}) \geqslant 0$ a.e.*

Property 4 *$|\mathbf{E}(X|\mathcal{A})| \leq \mathbf{E}(|X||\mathcal{A})$ a.e.*

Property 5 *$\mathbf{E}(aX + bY|\mathcal{A}) = a\mathbf{E}(X|\mathcal{A}) + b\mathbf{E}(Y|\mathcal{A})$ a.e. for any $a, b \in \mathbb{R}$.*

Property 6 *If $X_n \nearrow X$, then $\mathbf{E}(X_n|\mathcal{A}) \nearrow \mathbf{E}(X|\mathcal{A})$ a.e.*

Property 7 *If $\mathcal{A}_1 \subset \mathcal{A}_2 \subset \mathcal{F}$, then*

$$\mathbf{E}(X|\mathcal{A}_1) = \mathbf{E}(\mathbf{E}(X|\mathcal{A}_2)|\mathcal{A}_1) = \mathbf{E}(\mathbf{E}(X|\mathcal{A}_1)|\mathcal{A}_2) \text{ a.e.}$$

Proof Note that the variable $\mathbf{E}(X|\mathcal{A}_1)$ is $\mathcal{A}_1$ and $\mathcal{A}_2$-measurable because $\mathcal{A}_1 \subset \mathcal{A}_2$. By Property 2, we have

$$\mathbf{E}(\mathbf{E}(X|\mathcal{A}_1)|\mathcal{A}_2) = \mathbf{E}(X|\mathcal{A}_1).$$

8.4 Conditional Expectation

Now, let $A \in \mathcal{A}_1 \subset \mathcal{A}_2$. By the definition of conditional expectation, we have

$$\int_A \mathbf{E}(X|\mathcal{A}_2) \, d\mathbf{P} = \int_A X \, d\mathbf{P} = \int_A \mathbf{E}(X|\mathcal{A}_1) \, d\mathbf{P}.$$

Hence, $\mathbf{E}(\mathbf{E}(X|\mathcal{A}_2)|\mathcal{A}_1) = \mathbf{E}(X|\mathcal{A}_1)$, which ends the proof. □

Property 8 $\mathbf{E}X = \mathbf{E}(\mathbf{E}(X|\mathcal{A}))$ a.e.

Proof Let $\mathcal{A}_1 = \{\emptyset, \Omega\}$ and $\mathcal{A}_2 = \mathcal{A}$. Applying Properties 1 and 6, we obtain

$$\mathbf{E}X = \mathbf{E}(X|\mathcal{A}_1) = \mathbf{E}(\mathbf{E}(X|\mathcal{A})|\mathcal{A}_1) = \mathbf{E}\mathbf{E}(X|\mathcal{A}).$$

This formula is a generalization of the Total Probability Formula. □

Property 9 *If the variable X is independent of the σ-field $\mathcal{A}$, i.e.,*

$$\mathbf{P}(X^{-1}(B) \cap A) = \mathbf{P}(X^{-1}(B))\mathbf{P}(A) \qquad \text{for all } A \in \mathcal{A}, B \in \mathcal{B}(\mathbb{R}),$$

then $\mathbf{E}(X|\mathcal{A}) = \mathbf{E}X$ a.e.

Proof $\mathbf{E}X$ is $\mathcal{A}$-measurable, as is any constant function. Since X is independent of the σ-field $\mathcal{A}$, the variables X and $\mathbf{1}_A$ for $A \in \mathcal{A}$ are independent. Hence,

$$\int_A X \, d\mathbf{P} = \int X \mathbf{1}_A \, d\mathbf{P} = \int X \, d\mathbf{P} \int \mathbf{1}_A \, d\mathbf{P} = \int_A \mathbf{E}X \, d\mathbf{P}.$$

□

Property 10 *If Y is a bounded $\mathcal{A}$-measurable random variable, then*

$$\mathbf{E}(XY|\mathcal{A}) = Y\mathbf{E}(X|\mathcal{A}) \text{ a.e.}$$

Proof We can see that $Y\mathbf{E}(X|\mathcal{A})$ is $\mathcal{A}$-measurable. Assume that $Y = \mathbf{1}_B$ for some $B \in \mathcal{A}$. Then, for any $A \in \mathcal{A}$,

$$\int_A \mathbf{E}(X\mathbf{1}_B|\mathcal{A}) \, d\mathbf{P} = \int_A X\mathbf{1}_B \, d\mathbf{P} = \int_{A \cap B} X \, d\mathbf{P}$$

$$= \int_{A \cap B} \mathbf{E}(X|\mathcal{A}) \, d\mathbf{P} = \int_A \mathbf{1}_B \mathbf{E}(X|\mathcal{A}) \, d\mathbf{P}.$$

If Y is a simple function, i.e., $Y = \sum_{k=1}^{n} y_k \mathbf{1}_{B_k}$, $B_1, \ldots, B_n \in \mathcal{A}$, then

$$\mathbf{E}(XY|\mathcal{A}) = \sum_{k=1}^{n} y_k \mathbf{E}(X\mathbf{1}_{A_k}|\mathcal{A}) = \sum_{k=1}^{n} y_k \mathbf{1}_{A_k} \mathbf{E}(X|\mathcal{A}).$$

Now, we need to use the approximating sequence lemma and Property 6 to obtain the desired equality for any bounded, $\mathcal{A}$-measurable variable Y. □

Example 8.20 Assume that the σ-field $\mathcal{A}$ is atomic, i.e., $\mathcal{A} = \sigma\{A_1, A_2, \dots\}$, for a sequence of pairwise disjoint sets $A_1, A_2, \dots \in \mathcal{A}$ such that $\Omega = \bigcup_{k=1}^{\infty} A_k$. If $\mathbf{E}|X| < \infty$, then $\mathbf{E}(X|\mathcal{A})$, being an $\mathcal{A}$-measurable function, is of the following form:

$$\mathbf{E}(X|\mathcal{A}) = \sum_{k=1}^{\infty} x_k \mathbf{1}_{A_k}$$

for some constants $x_1, x_2, \dots \in \mathbb{R}$. To determine these constants, note that

$$\int_{A_n} \mathbf{E}(X|\mathcal{A}) \, d\mathbf{P} = \int_{A_n} x_n \mathbf{1}_{A_n} \, d\mathbf{P} = x_n \mathbf{P}(A_n).$$

At the same time, from the definition of conditional expectation, it follows that:

$$\int_{A_n} \mathbf{E}(X|\mathcal{A}) \, d\mathbf{P} = \int_{A_n} X \, d\mathbf{P} = \mathbf{E}(X \mathbf{1}_{A_n}).$$

Finally,

$$\mathbf{E}(X|\mathcal{A}) = \sum_{k=1}^{\infty} \frac{\mathbf{E}(X \mathbf{1}_{A_k})}{\mathbf{P}(A_k)} \mathbf{1}_{A_k}.$$

Definition 8.21 Let X and Y be random variables on $(\Omega, \mathcal{F}, \mathbf{P})$ and let $\mathbf{E}|X| < \infty$. The *conditional expectation of X given Y* is the random variable $\mathbf{E}(X|Y)$ defined by the formula

$$\mathbf{E}(X|Y) = \mathbf{E}(X|\sigma(Y)),$$

where $\sigma(Y) = \{Y^{-1}(B) : B \in \mathcal{B}\}$ is the σ-field generated by Y.

Theorem 8.22 *If X and Y are random variables on $(\Omega, \mathcal{F}, \mathbf{P})$ and $\mathbf{E}|X| < \infty$, then there exists a Borel function $h : \mathbb{R} \to \mathbb{R}$ such that*

$$\mathbf{E}(X|Y) = h(Y).$$

Proof We will prove that for any $\sigma(Y)$-measurable random variable Z, the function h exists. Assume first that $Z = \mathbf{1}_A$ for some $A \in \sigma(Y)$, i.e., $A = \{\omega : Y(\omega) \in B\}$ for some Borel set $A \in \mathcal{B}$. Then,

$$Z(\omega) = \mathbf{1}_A(\omega) = \mathbf{1}_B(Y(\omega)).$$

8.4 Conditional Expectation

Now, we can define $h = \mathbf{1}_B$. If Z is a simple variable, i.e., $Z = \sum_{k=1}^{n} z_k \mathbf{1}_{A_k}$, then, in a similar way, we get $Z(\omega) = h(Y(\omega))$ for $h(x) = \sum_{k=1}^{n} x_k \mathbf{1}_{B_k}(x)$.

If Z is a non-negative $\sigma(Y)$-measurable variable, then there exists a sequence of simple $\sigma(Y)$-measurable variables Z_n such that $Z_n(\omega) \nearrow Z(\omega)$. We know that for every simple variable Z_n, there is a Borel function h_n such that $Z_n(\omega) = h_n(Y(\omega))$. Therefore, $h_n(Y(\omega)) \to Z(\omega)$ for each $\omega \in \Omega$. Hence, the sequence of functions h_n has a limit at least on the set of values of the variable Y, and we can assume that

$$h(x) = \begin{cases} \lim_{n \to \infty} h_n(x) & \text{if this limit exists;} \\ 0, & \text{otherwise.} \end{cases}$$

Now, we just need to apply the obtained results to the variables Z^+ and Z^- which have separated supports and such that $Z = Z^+ - Z^-$. □

8.4.2 Exercises

367. We consider the probability space $([0, 1], \mathcal{B}([0, 1]), \lambda_1)$, where λ_1 is the Lebesgue measure on $[0, 1]$. Let $\mathcal{A} = \sigma\{[0, 1/3), \{1/3\}, (1/3, 2/3)\}$. Find the conditional expectation $\mathbf{E}(X|\mathcal{A})$ for the following random variables:

    (a) $X(\omega) = \omega$;
    (b) $X(\omega) = \sin(\pi\omega)$;
    (c) $X(\omega) = \omega^2$;
    (d) $X(\omega) = 1 - \omega$;
    (e) $X(\omega) = \begin{cases} 1 & \text{for } \omega \in [0, 1/3]; \\ 2 & \text{for } \omega \in [1/3, 1]. \end{cases}$

    Determine the distributions of the obtained random variables.

368. A random variable $X(\omega) = \omega$ is defined on the probability space $(\mathbb{R}, \mathcal{B}, \gamma)$, where γ is the exponential distribution $\Gamma(1, 1)$. Determine $\mathbf{E}(X|\mathcal{A})$ if $\mathcal{A} = \sigma\{[k, k+1) : k \in \mathbb{N}_0\}$.

369. A simple random variable X takes exactly n different values. Is it true that $\mathbf{E}(X|\mathcal{A})$ also takes at most n values?

370. Prove that if the σ-field $\mathcal{A}$ consists of events of probability 0 or 1, then $\mathbf{P}\{\mathbf{E}(X|\mathcal{A}) = \mathbf{E}X\} = 1$.

371. Assume that σ-fields $\mathcal{A}_1$ and $\mathcal{A}_2$ are independent, i.e., for any $A_1 \in \mathcal{A}_1$, $A_2 \in \mathcal{A}_2$ we have $\mathbf{P}(A_1 \cap A_2) = \mathbf{P}(A_1)\mathbf{P}(A_2)$. Prove that if $\mathbf{E}|X| < \infty$ and $\mathbf{E}|Y| < \infty$, then the random variables $\mathbf{E}(X|\mathcal{A}_1)$ and $\mathbf{E}(Y|\mathcal{A}_2)$ are independent.

372. Let X and Y be independent random variables with the same distribution and finite expectation. Prove that

$$\mathbf{E}(X|X+Y) = \mathbf{E}(Y|X+Y) = \frac{X+Y}{2} \quad \text{a.e.}$$

373. Does the random variable $\mathbf{E}(X|Y)$ have to be $\sigma(X)$-measurable?

374. Let $X_1, X_2, \ldots$ be independent random variables with the same distribution, $\mathbf{E}|X_1| < \infty$, $S_n = X_1 + \cdots + X_n$, $\mathcal{F}_n = \sigma(X_1, \ldots, X_n)$. Calculate $\mathbf{E}(S_k|\mathcal{F}_n)$.

375. Let $f(x, y)$ be the density function of a random vector (X, Y). Prove that the conditional expectation of Y given X is equal to $g(X)$, where

$$g(x) = \frac{\int_{\mathbb{R}} y f(x, y)\, dy}{\int_{\mathbb{R}} f(x, y)\, dy} = \int_{\mathbb{R}} y \, \frac{f(x, y)}{\int_{\mathbb{R}} f(x, y)\, dy}\, dy.$$

Hence, in a not very precise but useful way, we can say that the conditional expectation is equal to the expected value with respect to the conditional distribution.

Chapter 9
Hints and Solutions to the Exercises

Exercises 1.2.4

3. It is enough to use Newton's formula, also called the binomial formula: $(a+b)^n = \sum_{k=0}^{n} \binom{n}{k} a^k b^{n-k}$.
4. You can simply expand Newton's symbols and make use of Exercise 3. You can also differentiate the equality $(1+x)^n = \sum_{k=0}^{n} \binom{n}{k} x^k$ at the point $x=1$.
5. When answering the second question, remember that the players are numbered.
6. Note that the answer $\binom{13}{7}\binom{39}{6}$ is wrong. It only guarantees that Adam gets exactly seven spades, the other players have not been dealt any cards!
8. It all depends on whether we number the rooms or beds, or maybe both rooms and beds. If we number the beds, then there are $4! = 24$ possibilities.
9. Since this problem can be found in any high school probability textbook, we omit the answer. However, we should remember that people are distinguishable.
10. $2^7 - 2$.
11. The tailor's tape measure is 150 cm long, and Ewa could cut it in any of 149 places, hence $\binom{149}{2}$.
12. Why not hide these aces right away in your pocket and choose only the remaining cards?
13. Unfortunately, we have to deal with a large number of cases. Of course, there are 7 single-scoop desserts. There is already a problem with the two-scoops desserts, because there are 21 of them where the scoops are of different flavors, and there are 7 of those where both scoops are the same.
14. The first letter of a word can be any of the 31 letters of the alphabet, so can the second and the subsequent letters. Can you apply the principle of multiplication?
15. Perhaps it is better not to get too inquisitive and look only at the cases without chords, that is, we look at the cases where the swallows are not sitting one under the other. Then, of course we have 5^6 possibilities if the swallows are indistinguishable.

16. Unfortunately, the task has been awkwardly worded. Since the flags are to be three-colored, then we cannot paint the green-red-green flag and the task becomes trivial.
17. We can write four-digit numbers as $x = 1000a + 100b + 10c + d$, $a \geqslant 1$, $a, b, c, d \in \mathbb{N}$. a) answer: 333; b) x is divisible by 11 if and only if $a+c-b-d$ is divisible by 11. However, it is easier to use the periodicity of the remainder from dividing successive natural numbers by a fixed number; answer 819; c) note that the events *the first number is the sum of the remaining numbers* and *the fourth number is the sum of the others* are not disjoint, e.g. the event a00a is counted twice; d) answer: 9^4.
19. In the first case, the answer is 15 (we assume that the remainder must be non-zero) and, after the replacement, it is 18.
20. $\binom{15}{4} \cdot 2^4$.
21. We can distinguish the order of pairs, the order of children in a pair, or both the order of pairs and the order in a pair.
22. We believe that the coins only differ in denomination. Why not place the one-cent and the five-cent coins into separate sets of boxes, and then combine the coins from the boxes with the same number.
23. All you need to do is to select the drawers into which we want to put a ball.
24. Let's put one ball in each drawer. Only the remaining balls should be put in randomly.
25. Note that this is equivalent to arranging k indistinguishable balls in n numbered drawers. Zero components correspond to empty drawers.
26. The upper bound is easily derived from the previous exercise.

Exercises 2.1.1

27. No, but to see it you need to present a counterexample.
29. We need to show that any open set is equal to the countable union of open rectangles such that their defining numbers a, b, c, d are rational. There are countably many such rectangles. Now, for each point (x, y) of the open set U, we find a rectangle $P(x, y)$ with rational ends such that $(x, y) \in P(x, y) \subset U$. Then,

$$U \subset \bigcup_{(x,y) \in U} P(x, y) \subset U,$$

hence we get the equality, and the sum is countable because there are countable many such rectangles.

30. (a) and (b) Note that we also get a one-point set here.
    (d) Attention! Here the generator of the σ-field is the set of rational numbers, as in (a), where the generators were sets $[0, 2/3]$ and $[2/3, 1]$.

31. All σ-fields generated in cases (a), (b) and (c) are identical. Can the same be said about fields, rings, and σ-rings?
32. No! To see this, one must construct a countable sequence of finite sets, the sum of which is an infinite set and its complement is also an infinite set.

Exercises 2.2.1

34. (a) $A = B$; (b) $B = \Omega \setminus A$; (c) $A = B$; (d) $A \cap B = \emptyset$; but this still has to be proved.
35. (a) B; (b) $(A \cap B) \cup (A \cap C) \cup (B \cap C)$; (c) $(A \cap B) \cup (A \cap C) \cup (B \cap C)$; d) B; but the proof is necessary.
36. $A \cap B = \emptyset$.
37. They exist; you just need to choose $A \cap B \subset C$.
38. Let 0 denote the head and 1 denote the tail. The space Ω can be described in at least two ways:

$$\Omega = \{(0, 0, 0), (0, 0, 1), (0, 1, 0), \ldots, (1, 1, 1)\},$$

$$\Omega = \{(a_1, a_2, a_3) : a_i \in \{0, 1\}, i = 1, 2, 3\}.$$

39. First, we need to number the cards e.g., the order of the suits is as follows: one is an ace, two through ten are their numbers, eleven is a jack, twelve is a queen, thirteen is a king. The order of the suits: clubs, diamonds, hearts, spades. Thus card number 49 is the ten of spades.

Exercises 2.3.2

40. 2 and 2^n.
41. For 129 and 130 the answer is no. However, a justification is needed.
42. Just take the trivial σ-field in a space containing more than three elements.
46. It is enough to use the probability continuity theorem.
48. No! You need to give an example of a set $A \neq \emptyset$ for which $\mathbf{P}(A) = 0$.
49. You need to show that the subsets of the sets of measure zero from $\mathcal{F}_c$ also belong to $\mathcal{F}_c$.

Exercises 2.4.1

51. Of course $\{\emptyset, \Omega\}$ and 2^Ω. Which of them is complete?

52. Note that $\mathbf{P}(A) \geqslant \varepsilon\lambda(A)$ for every Borel set A. Consequently, if E were **P**-measurable, then we would have

$$\mathbf{P}([-1, 2]) = \mathbf{P}\left(\bigcup_{n=1}^{\infty}(E + q_n)\right) = \sum_{n=1}^{\infty}\mathbf{P}(E + q_n) \geqslant \varepsilon \sum_{n=1}^{\infty}\lambda(E + q_n) = \infty.$$

53. We need to show that the Lebesgue measure of this set is equal to zero, and since it is a countable sum of one-point sets, it suffices to show that the Lebesgue measure of a point is equal to zero.

Exercises 2.5.1

54. The number of all irreducible fractions is equal to $2(5 \cdot 3 + 3 \cdot 7)$.
55. For example, for $k = 3$, we have $n(\Omega) = \binom{90}{5}$, $n(A_k) = \binom{87}{2}$, where A_k means that among the numbers drawn are all k numbers that the player has bet on.
56. There are only four such stones.
57. 6^{1-n}.
58. It is worth noting that the two possible approaches to this task lead to the same result. If we can't distinguish between the people, then $p = \binom{n+k-m}{n-m}/\binom{n+k}{n}$. If we distinguish between them, then $n(A) = \binom{n}{m}m!(n + k - m)\ldots(k + 1)$, but the number of events in Ω must also be counted differently.
59. The cardinality of the set of favorable events is: $\binom{2n-2}{n-2} + \binom{2n-2}{n}$.
60. In the first case, the probability of drawing a white ball is equal to 0.5 in each draw. We are not interested in the order in which the balls are selected, so we identify all sequences with exactly n white balls—there are $\binom{2n}{n}$ of them. Now, it suffices to note that each such sequence occurs with a probability of 2^{-2n}. The second case, when we draw without returning the balls, is easier.
61. $\binom{25}{5}\binom{20}{5}\binom{15}{5}\binom{10}{5}\binom{5}{5} \cdot 5^{-25}$.
62. Let k be the number of drawn balls. Consider all k-element sequences of zeros (white balls) and ones (black balls) that have a zero in the first position. If we write this sequence from the end, we get a sequence with a zero in the last position. Hence, the number of elementary events favoring both events is the same.
63. $(1 + \alpha)^{-1}$.
64. Suppose there are n white balls and k black balls. Without loss of generality, it can be assumed that $k \leqslant n$. Then, the sought probability is:

$$p = \frac{n^2 + k^2}{(n + k)^2} = \frac{1 + x^2}{(1 + x)^2} =: f(x),$$

where $x = k/n \in (0, 1]$. It is easy to show that f is a decreasing function and $f(1) = 0.5$. How will the function f change when the draws are done without return?

65.

67.
$$\text{(a)} \ \frac{\binom{N}{n}}{\binom{N+n-1}{n}}, \quad \text{(b)} \ \frac{\binom{N}{n}n!}{N^n}.$$

$$p(k) = \frac{\binom{30}{k}7^{30-k}}{8^{30}}.$$

68. When counting the cardinality of the event A, we can ignore the piggy-bank. Of course, we treat coins as indistinguishable.
69. $n(A) = n(n+1)(n^2 - n + 3)/6$.
70. $n(A) = n(n-1)(n-2)/6$.
71. $\binom{2n-1}{n}/\binom{3n-1}{2n}$.
72. Let us divide both letters and envelopes into even and odd numbered ones. The sum will be even if the sheet and envelope have even numbers or both have odd numbers. Hence, the cardinality of the event we are interested in is equal to $(n!)^2$.
73. Only item (c) may cause some difficulties. First, note that the sequence from the first ace to the last one consists of $3k + 4$ cards, so $3k + 4 \leq 52$. Now, choose the position of the first ace in the deck ($52 - 3k - 4 + 1$ ways), place the aces in the already established positions (4! ways), then shuffle the rest of the cards and place them in the empty places.
75. First, let's set up the 6 rooks according to the set rules. We will have four empty spaces left, two of which belong to the forbidden diagonal, so there is only one way to arrange the last two rooks (two ways if we are considering numbered rooks, but then the calculations are more difficult!).
76. It is easier to calculate the probability of having at least k such pairs, and then calculate $\mathbf{P}\{X = k\} = \mathbf{P}\{X \geq k\} - \mathbf{P}\{X \geq k+1\}$.
79. A correct description of the set Ω is only a half of the solution. Let x be the bridge chosen by the person leaving Burghers' Island and y the bridge chosen by the other. Then, we have $\Omega = \{(x, y) \colon x, y \in \{1, \ldots, 7\}\}$.
80. First, choose k pairs of shoes, and then choose one shoe from each of the selected pairs (the shoes in a pair are different: left and right!).
81. Planets and moons are distinguishable. You should first number the planets (e.g., in order of distance from the sun), then number the moons and add the moons to the planets. We agree that if a planet has received the moons numbered k and ℓ, $k < \ell$, then the k-th moon is closer to the planet than the ℓ-th moon. Hence, $n(\Omega) = 4! \cdot 5! \cdot 4^5$, $n(A) = 4! \cdot 4 \cdot 5! \cdot 3^5$.

82. Electrons and atomic nuclei should be treated as indistinguishable elements. Let us arrange five indistinguishable balls (electrons) into four unnumbered boxes (atomic nuclei). Then, we get:

$$5 = 5 + 0 + 0 + 0 = 4 + 1 + 0 + 0 \ldots$$

Exercises 2.6.1

84. $\Omega = [0, 5]$, $A = (1 + \sqrt{3}, 5]$.
86. $5/9$.
87. $\Omega = [0, \ell] \times [0, \ell]$, $P(A) = 1/4$.
88. Only the distance from the center of the coin to the nearest line below this center is important, so $\Omega = [0, \ell)$.
89. $\Omega = [-1, 1]^2$, $A = \{(p, q) \in \Omega: p^2 - 4q > 0\}$.
90. $\Omega = [a, b]^3$, $A = \{(x, y, z) \in \Omega: x \leqslant z \leqslant y \text{ or } y \leqslant z \leqslant x\}$.
91. If we define success properly, then it is enough to calculate the probability of getting k successes in n trials.
92. This is also the problem of getting n successes.
93. Difficult! $\Omega = K^2$, and in the space $\mathbb{R}^4$ it is hard to see anything! However, determining the position of the point $A = (x, y)$, we will get the cut D_A of the sought set $D \subset K^2$ by the plane $(x, y, z, t): (z, t) \in K$. The answer for $K = [0, 1]^2$ is:

$$\int_0^1 \int_0^1 \left(1 - (x - y)^2 - (1 - x - y)^2 \mathbf{1}(x + y < 1)\right) dx\, dy.$$

94. For those who do not remember the formula for the area of a circular segment, let us recall that the definite integral over an interval is equal to the area under the integrated curve.
95. The length of the curve described by the function $f(x)$ on the segment $[a, b]$ is given by the formula

$$\int_a^b \sqrt{1 + (f'(x))^2}\, dx.$$

Exercises 2.7.1

101. Yes.
102. No.

9 Hints and Solutions to the Exercises

103. $\frac{1}{3}\left(\frac{\alpha}{\alpha+1} + \frac{\beta}{\beta+1} + \frac{\gamma}{\gamma+1}\right).$
108. $\frac{N-2}{N+M-2}.$
109. $2/3.$
110. $\mathbf{P}(5) = \mathbf{P}(4) = 1/3.$
111. $\frac{507}{1210}.$
112. Let us choose two of the three baskets. With a probability of 2/3, we get the baskets of various kinds. If we now buy 1.5 kgs of nuts from each of them, how many wormy nuts will we buy?

Exercises 2.8.1

120. The events are pairwise independent but not independent.
121. The events $A \cap B$, $A \cap C$, $B \cap C$ are pairwise independent but not independent.
124. These events are not even pairwise independent.
125. $\frac{9}{19}.$
126. $2^n - 1 - n.$

Exercises 2.9.3

127. We have five problems, so five experiments. The probability of success is the probability that the topic has been discussed and that Adam is able to solve the problem. Hence, $p = 0.736$. The task contains redundant and therefore confusing information that there were supposed to be 15 lectures.
128. The problem becomes easy if we consider n even and n odd separately.
129. (a) $n \cdot \frac{1}{6}\left(\frac{5}{6}\right)^{n-1}$, (b) $1 - \left(\frac{5}{6}\right)^n.$
130. The event A.
131. The events are equally probable.
132. Recall that $\sum_{k=0}^{n}\left(\binom{n}{k}\right)^2 = \binom{2n}{n}.$
134. $n \geqslant 9.$
135. Approximately 0.9033.
136. $1 - (\lambda p + 1 - p)^n.$
137. The solution of this problem for $n = 7$ with a full justification of the strategy used can be easily found in the literature on Hamming codes, as well as in the popular science literature. However, you are advised to try and find this solution yourself—maybe you can find a simpler and therefore nicer solution?

Exercises 2.10.1

143. $\mathbf{P}(A) = 0$.
144. $\mathbf{P}(A) = 1$.

Exercises 3.1.1

148. Only Y and Z are random variables.

Exercises 3.2.1

154. These variables have identical distributions. However, $\mathbf{P}\{\omega\colon X(\omega) \neq Y(\omega)\} = 1$, which means that they are different everywhere.
155. (a) $\{(a,b,c)\colon a = 0, 0 \leq b \leq 1-c, 0 \leq c \leq 1\}$, (b) $\{(a,b)\colon b = 0, 0 \leq a \leq 1\}$ (left-continuity property).
156. Just to verify: the cumulative distribution function of X has a jump of height $1/3$ at the point $t = 3$, and the cumulative distribution function of Y is continuous but not a piecewise linear function.
157. (c) $F(t) = \mathbf{1}_{(0,\infty)}(t)$.
158. Hint: Calculations must be performed separately for $a > 0$, $a < 0$ and $a = 0$.
159. Suppose that it does not have to be the case, and let F be a cumulative distribution function with uncountably many discontinuity points. Let A_n denote the set of those points where the distribution function jumps exceed $1/n$. If all sets A_n are finite, then the cumulative distribution function has countably many of these points, so at least one of these sets, for example A_{n_0}, is infinite. It should not be difficult to derive from this that the considered distribution function is unbounded.

Exercises 3.3.9

165. $[(r-1)/p] + 1$.
166. $\mathbf{P}(X = k) = \binom{52}{13}^{-1}\binom{4}{k}\binom{52-k}{13-k}$, $k = 0, 1, 2, 3, 4$.
168. $5/16$.
169. The Bernoulli distribution with parameters $n = 10$, $p = 0.5$.
170. Careful here! This is not the distribution of the waiting time for the third success, but the arithmetic mean of two such distributions. The result would be different if he randomly determined whether he should roll a die or toss a coin before each move.

Exercises 3.4.7

171. The exponential distribution with parameter λ^2.
172. The gamma distribution with parameters b^2, p.
173. The Cauchy distribution with parameters $1, 0$.
179. $N(am+b, |a|\sigma)$.
180. $f(x) = 2\lambda x e^{-\lambda x^2} \mathbf{1}(x > 0)$.
181. Determine the length of the third side $X = h(\cos\varphi)$ from the Law of Cosines, where the variable φ has the uniform distribution on $[0, 2\pi)$.
182. $\Omega = \{(x, y): x, y \in [0, 1]\}$.
 $X(x, y) = \sqrt{x^2 + y^2}\mathbf{1}(x^2 + y^2 \leqslant 1) + \min\{x, y, 1-x, 1-y\}\mathbf{1}(x^2 + y^2 > 1)$.

Exercises 3.6.1

184. You can take X^3 as "any other" variable.
186. Let F be the cumulative distribution function with density f. Now, just calculate the appropriate distribution functions of the variables Z and U, and then differentiate them. For example, $\mathbf{P}\{Z < u\} = \mathbf{P}\{X < u, Y < u\} = F^2(u)$. Hence, $f_Z(u) = 2F(u)f(u)$.
187. To calculate the distribution of U_2, note that

$$\mathbf{P}\{U_2 < u\} = 3\mathbf{P}\{X < u, Y < u, Z \geqslant u\}.$$

188. $F_Z(t) = F(t)G(t)$, $F_W(t) = F(t) + G(t) - F(t)G(t)$, $F_T(t) = F(t/2)G(t)$, $F_U(t) = 1 - (1 - F(\sqrt[3]{t}))(1 - G(t))$.
188. Find the cumulative distribution functions of the variables $Y_n = \sum_{i=1}^n \frac{X_i}{2^i}$ first.
195. Note that $\sum_{k=1}^n X_k$ has distribution $\Gamma(n, a)$. To calculate the distribution function of Y, we use the total probability formula: for $u > 0$,

$$\mathbf{P}\{Y < u\} = \sum_{k=1}^\infty pq^{n-1} \int_0^u \frac{a^n}{(n-1)!} x^{n-1} e^{-ax} \, dx.$$

196.

$$f(z) = \frac{1}{b-a}\Big[\Phi(z-a) - \Phi(z-b)\Big].$$

197. $f_Z(x) = 2\pi^{-2} x(\sinh(x))^{-1}$.

Exercises 3.7.1

198. X and Y are not independent.
200. Recall that variables of continuous type are independent if their joint density is the product of the density functions of separated variables.
201. Yes, it is possible. Just remember that the integral of the density function is equal to one.
203. (b) $F_T(t) = 1 - e^{-\alpha t} - \alpha t e^{-\alpha t}$, $t > 0$; (d) $F_V(t) = 1 - e^{-\alpha t} + e^{-3\alpha t}$; (f) $F_Z(t) = \frac{t}{t+1}, t > 0$.
204. (a) $\frac{p}{\pi} \frac{a}{a^2+(t-m)^2} + \frac{1-p}{\pi} \frac{a}{a^2+(t+m)^2}$; (b) $\lambda p e^{-\lambda t} I(t > 0) = \lambda(1-p) e^{\lambda t} I(t < 0)$.
205. Try to express the density as an integral, e.g., $f_Z(z) = \int_\mathbb{R} f(x, z-x) \, dx$, by calculating the corresponding distribution function first.
206. A lot of cases. If $u \in (0, 1)$, then

$$P\{Z < u\} = P\{X > u(X+Y), X+Y > 0\}$$
$$+ P\{X < u(X+Y), X+Y < 0\}$$
$$= \int_0^\infty \int_{-y}^{uy/(1-u)} f(x, y) \, dx \, dy + \int_{-\infty}^0 \int_{uy/(1-u)}^{-y} f(x, y) \, dx \, dy.$$

207. $\frac{1}{2}\left(1 + e^{-4t} - 2e^{-2t}\right)$.
208. $P\{X > Y\} = 0.5$.
209. Recall that there are two discrete geometrical distributions; one starts from zero, the other starts from one. Hence,

$$P\{Z_0 = k\} = (k+1)p^2 q^k, \ k = 0, 1, \ldots,$$
$$P\{Z_1 = k\} = (k-1)p^2 q^{k-2}, \ k = 2, 3, \ldots$$

The variable Z_1 has the Pascal distribution with parameter $r = 2$.
210. The variable Z has the Poisson distribution with parameter $\lambda_1 + \lambda_2$.
211. $e^{-\lambda p}(\lambda p)^n / n!$.
212. $\sqrt{p_1} + \sqrt{p_3} = 1$, $p_2 = 2\sqrt{p_1 p_3}$.
217. $\Gamma(p+q, a)$.
220. It is sufficient to show that the joint distribution function and/or the density of the vector (Z, W) is a function of separated variables. Nevertheless, it is worth finding the densities of the variables Z and W. For $u, v > 0$, we have

$$F_{Z,W}(u, v) = P\{Z < u, W < v\} = \int \cdots \int_{x+y<u, x/y<v} f_X(x) f_Y(y) \, dx \, dy$$
$$= \frac{v}{1+v}\left(1 - e^{-u}(1+u)\right);$$

$$f_{Z,W}(u,v) = \frac{1}{(1+v)^2} \cdot u e^{-u}.$$

221. Though slightly laborious, it is possible to calculate the following:

$$F_{Z,W}(u,v) = \mathbf{P}\{Z < u, W < v\} = \int \cdots \int_{x^2+y^2<u, x/y<v} \frac{1}{2\pi} e^{-(x^2+y^2)/2} \, dx \, dy$$

$$= \frac{1}{\pi}\left(\pi - \arctan\frac{1}{v}\right)\frac{1}{2}\left(1 - e^{-u/2}\right);$$

$$f_{Z,W}(u,v) = \frac{1}{\pi(1+v^2)} \cdot \frac{1}{2} e^{-u/2}.$$

222. For $u, v > 0$, we have

$$F_{Z,W}(u,v) = \frac{a^{p+q}}{\Gamma(p)\Gamma(q)} \int_0^{\frac{uv}{1+v}} x^{p-1} e^{-ax} \int_{x/v}^{u-v} y^{q-1} e^{-ay} \, dy \, dx$$

$$f_{Z,W}(u,v) = \frac{a^{p+q}}{\Gamma(p+q)} u^{p+q-1} e^{-au} \cdot \frac{v^q(1+v)^{-(p+q)}}{B(p,q)}.$$

Exercises 4.1.1

225. The game is fair if it has a win expectation equal to zero, so $a = 2b$.

226. Without losing generality, we can assume that a_1 is the largest number. Then,

$$\sqrt[n]{\mathbf{E}X^n} = a_1 \sqrt[n]{p_1 + p_2(a_2/a_1)^n + \cdots + p_n(a_n/a_1)^n}$$
$$\leqslant a_1 \sqrt[n]{p_1 + \cdots + p_n} = a_1.$$

On the other hand, $\sqrt[n]{\mathbf{E}X^n} \geqslant a_1 \sqrt[n]{p_1} \to 1$. The second limit is even easier to find.

228. Note that the variable X has a hypergeometric distribution with parameters $N, M, k, k \leqslant N, k \leqslant M$:

$$\mathbf{P}\{X = \ell\} = \frac{\binom{N}{\ell}\binom{M}{k-\ell}}{\binom{N+M}{k}}, \qquad \ell = 0, 1, \ldots k.$$

In particular, this means that

$$\sum_{\ell=0}^{k} \binom{N}{\ell}\binom{M}{k-\ell} = \binom{N+M}{k}.$$

This identity shall be used in calculating $\mathbf{E}X$.

Exercises 4.2.1

231. Of course, these variables cannot be independent!
235. Consider the random variables $Y_i = X_i/(X_1 + \cdots + X_n)$. Of course, they have the same distributions, as well as the same expected values. Now, just note that $\mathbf{E}(Y_1 + \ldots Y_n) = \mathbf{E}1 = 1$.
236. $\mathbf{E}\max\{0, X\} = 0.5$.
237. Let $x = \mathbf{E}X^+ \geq 0$, $y = \mathbf{E}X^- \geq 0$. Then, $a = x - y$, $b = x + y$ and it is enough to draw conclusions.
238. $\mathbf{P}\{X \geq k\} = \sum_{n=k}^{\infty} \mathbf{P}\{X = n\}$, $k \in \mathbb{N}$. Now, it is enough to change the order of summation in the expectation formula.
239. This is a simple conclusion from Exercise 238.

Exercises 4.3.1

240. The appropriate integral should be written as the sum of the integrals over the sets $\{\omega: |X(\omega)| \leq 1\}$ and $\{\omega: |X(\omega)| > 1\}$. On the first set, $|X(\omega)| \leq 1$, and on the second, $|X(\omega)| \leq X^2(\omega)$. Thus, $\mathbf{E}|X| \leq \mathbf{E}X^2 + \mathbf{P}\{\omega: |X(\omega)| \leq 1\}$.

Exercises 4.4.1

248. It does not exist.
249. We do not need to integrate anything. Just remember that the parameter p in both distributions can have any positive value, and that the integral of the distribution density is equal to 1.
250. $\Omega = [0, 297]$, $\mathbf{P}(d\omega) = d\omega/297$. Answer: $\frac{3}{4} \cdot 210 \cdot 297$.

9 Hints and Solutions to the Exercises

Exercises 4.5.1

251. Just apply integration by parts.
252. This is a straightforward conclusion from Exercise 216.
253. Attention! These cumulative distribution functions can be discontinuous.
254. It is enough to note that C is the set of real numbers which contain no 1's in their ternary (base three) representation.

Exercises 5.1.1

260. (b) This is a simple application of the fact that the derivative of an absolutely convergent power series is equal to the series of derivatives of its components.
266. This exercise, when it appears in the next chapter, will be even easier.
267. Note that the function $f(a) = \mathbf{E}(X - a)^2$ is quadratic, so its minimum is easy to determine.
268. Note that $\text{Var}(XY) = \text{Var}X \cdot \text{Var}Y + \text{Var}X (\mathbf{E}Y)^2 + \text{Var}Y (\mathbf{E}X)^2$.
270. We already know the value of $\mathbf{E}X$.

Exercises 5.2.1

274. $\mathbf{P}\{15 \leqslant X \leqslant 45\} \geqslant 0.9$.
275. $n \geqslant 2560$.

Exercises 5.3.3

281. $|\rho(Z, W)| = |\alpha|$.
282. $\rho(X, Y) = 0$, but the variables are not independent.
284. $\mathbf{P}\{X > \frac{1}{2}, Y > \frac{1}{2}\} \neq \mathbf{P}\{X > \frac{1}{2}\}\mathbf{P}\{Y > \frac{1}{2}\}$.
286. To determine the distribution function of the variable X, let us recall that the area of the side surface of the solid formed by the rotation of the curve $y = f(x)$, $x \in [a, b]$ around the axis OX is equal to $2\pi \int_a^b f\sqrt{1 + (f')^2}\, dx$. Answer: X has a uniform distribution, the density of the vector (X, Y) at the point (x, y) is a function that depends on $x^2 + y^2$.

Exercises 5.4.1

294. It is worth noticing that for $\mathrm{Cov}(X, Y) = 0$, the joint density for the vector (X, Y) is a function with separated variables.
295. $\mathbf{E}Y = \mathbf{0}$, $\mathrm{Var}\,X = a\Sigma a^T$, where $a = (a_1, \ldots, a_n)$.
296. The final calculations are as follows

$$\frac{1}{2\pi}\int_{\mathbb{R}} e^{-y^2/2}\int_y^\infty e^{-(x-y)^2/2}\mathrm{d}x\,\mathrm{d}y = \frac{1}{\sqrt{2\pi}}\int_{\mathbb{R}} e^{-y^2/2}\int_0^\infty e^{-r^2/2}\,\mathrm{d}r\,\mathrm{d}y = \frac{1}{2}.$$

297. (b) Let A_φ be the arc of the unit circle between the positive semi-axis OX and the line $y = \mathrm{tg}(\varphi)x$. Then, $\mathbf{P}\{U \in A_\varphi\} = \mathbf{P}\{X \geqslant 0, \mathrm{tg}(\varphi)X \geqslant Y \geqslant 0\}$. c) It is enough to calculate $\mathbf{P}\{R \in [0, r], U \in A_\varphi\}$ for any $r > 0$, $\varphi \in [0, 2\pi)$.
298. First of all, you need to make sure that this function *is* a density function.

Exercises 6.1.1

310. $\varphi_Y(t) = (1 + it/b)^{-(p_1+\cdots+p_n)}$.
311. This can be demonstrated by considering a convex combination of the distributions of these variables. But it can be done a bit differently: we define a random variable Θ, independent of $X_1, \ldots, X_n$, which takes the values $1, \ldots, n$ with probabilities $p_1, \ldots, p_n$, respectively. Now, simply note that φ is the characteristic function of the variable

$$Y = \sum_1^n X_k \mathbf{1}_{\{\Theta=k\}}.$$

312. Let $X_1, X_2, \ldots$ be independent random variables with distribution function F, and let Θ be a random variable with a geometric distribution with parameter $\frac{1}{2}$ independent of all X_i. Now, you only need to calculate the characteristic functions of the following variables: $\sum_1^\Theta X_i$, $X_1\mathbf{1}_{\{\Theta=1\}} - X_2\mathbf{1}_{\{\Theta>1\}}$, $X_1 - X_2$.
313. Suppose that φ is the characteristic function of X and let θ be a variable independent of X with the uniform distribution on $[0, 1]$. Now, you only need to calculate the characteristic function of the variable $Y = X\theta$.

Exercises 6.2.1

316. Assume that $a = 0$. Since $\mathbf{P}\{X = 0\} = p$, there exists a probability measure μ such that $\mathbf{P}_X = p\delta_0 + (1-p)\mu$. Hence, $|\varphi(t) - p| = (1-p)|\widehat{\mu}(t)| \leqslant (1-p)$, so the smallest possible distance φ from the OY axis is $p - (1-p) = 2p - 1 > 0$.

If $a \neq 0$ we get $|\varphi(t) - pe^{ita}| = |\varphi(t)e^{-ita} - p| \leq 1 - p$, which does not change the distance of φ from the OY axis.

317. Yes, it may, but it is not a very random variable.
318. Just use Theorem 6.13.
319. This is another application of Theorem 6.13.
321. (a) $\varphi_{X_1}(a,b) = \mathbf{E}e^{i\langle(a,b),X_1\rangle} = (e^{ia} + e^{-ia} + e^{ib} + e^{-ib})/4 = 1 - \frac{a^2+b^2}{2} + o(a^2+b^2)$. (b) $\varphi_{X_1}(a,b) = 1 - \frac{a^2+b^2}{3} + o(a^2+b^2)$. (c) $\varphi_{X_1}(a,b) = 1 - \frac{2a^2+(b-a)^2}{6} + o(a^2+b^2)$.

We can now find the limit of $\varphi_{S_n/n}(a,b)$.

Exercises 6.3.1

324. Note that
$$\binom{n}{k}p_n^k(1-p_n)^{n-k} = \frac{(np_n)^k}{k!}\frac{n(n-1)\ldots(n-k+1)}{n^k}(1-p_n)^{-k}\left(1-\frac{np_n}{n}\right)^n.$$

325. Yes, the limit distribution is a Cauchy distribution with parameters mA, aA if X_i have the Cauchy distribution with parameters m, a.
326. You need to use Lemma 6.17. We are looking for a compact interval of the form $[-2/u, 2/u]$, where u should be selected similarly to Step 1 of the proof of the Lévy–Cramér theorem, but uniformly for all φ_n.
328. Note that
$$\binom{n}{k}p_n^k(1-p_n)^{n-k} = \frac{(np_n)^k}{k!}\frac{n(n-1)\ldots(n-k+1)}{n^k}(1-p_n)^{-k}\left(1-\frac{np_n}{n}\right)^n.$$

Exercises 6.4.1

334. Too difficult for you? That's good! The function $f(t) = \int_0^{\pi/2} \cos(t\cos x)\,dx$ is a special function—a Bessel function of the first kind. However, it is possible to show elementarily that $\varphi_{(X,Y)}(a,b) = \varphi_{(X,Y)}(\sqrt{a^2+b^2}, 0)$.

Exercises 7.2.1

341. (a) Note that $\mathbf{P}\{\sum^n X_k \geq 2^n\} \geq \mathbf{P}\{X_n = 2^n, X_{n-1} = 2^{n-1}\} = \frac{1}{4}$ because $2^k - 2 \geq |\sum_1^{k-1}(\pm 2^j)|$. Hence, it already follows that neither the Weak

nor the Strong Law of Large Numbers does not hold. (b) Yes. (c) Let us first show that random variables $\sum_1^n X_k$ are symmetric, e.g. by showing that their characteristic functions are real. Now simply note that $\mathbf{P}\{\sum_1^n X_k \geq n\} \geq \mathbf{P}\{X_n = n, \sum_1^{n-1} X_k \geq 0\} \geq \frac{1}{2}\frac{1}{2}$ and draw conclusions. (f) The Strong Law of Large Numbers holds. What about the Weak?

342. We have $\mathbf{E}X_n = 0$ and $\mathrm{Var}\, X_n = \ln n$, thus

$$n^{-2}\sum_{k=1}^n \mathrm{Var}\, X_n = n^{-2}\sum_{k=1}^n \ln n \leq n^{-1}\ln n \to 0 \quad \text{if} \quad n \to \infty.$$

343. Note that

$$\mathbf{P}\left\{\left|\frac{1}{n}\sum_{k=1}^n X_k\right| < \varepsilon\right\} = \frac{1}{2}\mathbf{P}\left\{\left|\frac{1}{n}\sum_{k=1}^{n-1} X_k + \alpha_n\right| < \varepsilon\right\}$$
$$+ \frac{1}{2}\mathbf{P}\left\{\left|\frac{1}{n}\sum_{k=1}^{n-1} X_k - \alpha_n\right| < \varepsilon\right\}.$$

Let $\varepsilon \in (0, \alpha)$. If $\omega \in \Omega$ such that $\left|\frac{1}{n}\sum_{k=1}^{n-1} X_k + \alpha_n\right| < \varepsilon$, then

$$\left|\frac{1}{n}\sum_{k=1}^{n-1} X_k - \alpha_n\right| = \left|\frac{1}{n}\sum_{k=1}^{n-1} X_k + \alpha_n - 2\alpha_n\right| > 2\alpha_n - \varepsilon > \alpha.$$

By changing α_n to $-\alpha_n$ in the above reasoning, we get that the sets under consideration are disjoint. Hence, $\mathbf{P}\left\{\frac{1}{n}\left|\sum_{k=1}^n X_k\right| < \varepsilon\right\} \leq \frac{1}{2}$ for every $n \in \mathbb{N}$ and it cannot converge to one.

344. Note first that

$$\mathrm{Cov}(X_i, X_j) \leq \sqrt{\mathrm{Var}\, X_i\, \mathrm{Var}\, X_j} \leq \mathrm{Var}\, X_i + \mathrm{Var}\, X_j.$$

Hence,

$$\mathrm{Var}\left(\sum_{k=1}^n X_k\right) = \sum_{k=1}^n \mathrm{Var}\, X_k + 2\Big(\mathrm{Cov}(X_1, X_2) + \cdots + \mathrm{Cov}(X_{n-1}, X_n)\Big)$$
$$= \sum_{k=1}^n \mathrm{Var}\, X_k + 2\left(\sum_{k=2}^n \mathrm{Var}\, X_{k-1} + \sum_{k=1}^n \mathrm{Var}\, X_{k+1}\right) \leq 5\sum_{k=1}^{n+1}\mathrm{Var}\, X_k.$$

Now we should use the assumption and prove the result.

9 Hints and Solutions to the Exercises

Exercises 7.3.1

346. Approximately 0.0793.
347. Obviously, X is the waiting time for the hundredth success, so we know the exact distribution. Let $X_i = 1$ if i-th passer-by buys a newspaper, otherwise $X_i = 0$. The variables X_i are independent, $\mathbf{E}X_i = 1/3$, $\mathrm{Var}\,X_i = 2/9$. We are looking for an estimate of the cumulative distribution function:

$$\mathbf{P}\{X < 100+n\} = \mathbf{P}\left\{\sum_1^{100+n} X_i > 100\right\} \sim 1 - \Phi\left(\frac{200-n}{\sqrt{2(100+n)}}\right).$$

348. Let X_i denote the error of the i-th approximation, $n = 1200$.

$$\mathbf{P}\left\{\left|\sum_{i}^{n} X_i\right| > 10\right\} \sim 2(1 - \Phi(1)) \sim 0.3174.$$

349. Let $m = \mathbf{E}X_k$, $\sigma^2 = \mathrm{Var}\,X_k$. Then

$$\mathbf{P}\left\{a < \sum_{k=1}^{n} X_k < b\right\} \sim \Phi\left(\frac{b-nm}{\sigma\sqrt{n}}\right) - \Phi\left(\frac{b-nm}{\sigma\sqrt{n}}\right).$$

350. This limit is equal to $\frac{1}{2}$ if $\mathbf{E}X_k = 0$.
351. If $x \leqslant 0$ the result is trivial. For $x > 0$, the probability is approximately equal to $2\Phi(xn^{\alpha-\frac{1}{2}}/\sigma) - 1$.
352. After applying the Central Limit Theorem, the condition can be written as $\Phi(\sqrt{3}(2a_n - \sqrt{n})) \to p$.
353. Note that

$$\frac{1}{\sqrt{n}}\sum_{k=1}^{n} X_k = \frac{1}{\sqrt{n}}\sum_{k=1}^{n}(X_k - \mathbf{E}X_k) + \frac{1}{\sqrt{n}}\sum_{k=1}^{n}\mathbf{E}(X_k - [X_k])$$

$$= \frac{1}{\sqrt{n}}\sum_{k=1}^{n}(X_k - \mathbf{E}X_k) + \frac{1}{\sqrt{n}}\sum_{k=1}^{n}\mathbf{E}\{X_k\}$$

$$= \frac{1}{\sqrt{n}}\sum_{k=1}^{n}(X_k - \mathbf{E}X_k) + \sqrt{n}\,\mathbf{E}\{X_1\}.$$

Now all you need to do is apply the Central Limit Theorem.

354. If $X_1, X_2, \ldots$ are independent and identically distributed with the Poisson distribution with parameter 1, then $\sum_{k=1}^n X_k$ has the Poisson distribution with parameter n. Hence,

$$e^{-n} \sum_{k=0}^n \frac{n^k}{k!} = \sum_{k=0}^n \mathbf{P}\left\{\sum_{j=1}^n X_j = k\right\} = \mathbf{P}\left\{0 \leqslant \sum_{j=1}^n X_j \leqslant n\right\}$$

$$= \mathbf{P}\left\{-\sqrt{n} \leqslant \frac{\sum_{j=1}^n X_j - n}{\sqrt{n}} \leqslant 0\right\} \stackrel{CTG}{\approx} \Phi(0) - \Phi(-\sqrt{n}) \to \frac{1}{2}.$$

Exercises 8.3.1

362. (a) We are looking for a non-negative function h, such that for any Borel set $A \subset [0, \infty)$, the following equality holds:

$$\int_A h(x) \frac{1}{\sqrt{2\pi}} e^{-x^2/2} \, dx = \int_A \frac{b^p}{\Gamma(p)} x^{p-1} e^{-bx} \, dx.$$

364. It is not true that if a distribution μ is absolutely continuous with respect to the measure λ, then the implication $(\lambda(A) > 0) \Rightarrow (\mu(A) > 0)$ holds. Hence, the support of μ does not have to be the entire real line, and supports of the measures μ and ν can be disjoint.

366. No, it isn't, but it should be proven.

Exercises 8.4.2

367. We see that $\mathcal{A}$ is an atomic σ-field, so we proceed as in Example 8.20.
368. Note that $\mathbf{P}(d\omega) = e^{-\omega} d\omega$ on the positive part and $\mathbf{P}(d\omega) = 0$ on the negative part of the real axis.
369. No, the number of values can increase. As a counterexample, it is sufficient to consider a two-valued random variable and a σ-field generated by three atoms.
370. Of course, the constant $\mathbf{E}X$ is an $\mathcal{A}$-measurable function. We still need to verify that the integration condition holds.

371. Let us go further, namely, if X is $\mathcal{A}_1$ measurable and Y is $\mathcal{A}_2$ measurable, then X, Y are independent. We start with the case when the variables are simple, i.e., $X = \sum x_i \mathbf{1}_{A_i}$, $Y = \sum y_j \mathbf{1}_{B_j}$. Then, for any Borel sets $C, D \subset \mathbb{R}$, we have:

$$\mathbf{P}\{X \in C, Y \in D\} = \mathbf{P}\Big(\bigcup_{x_i \in C} \bigcup_{y_j \in D} (A_i \cap B_j)\Big) = \sum_{x_i \in C} \sum_{y_j \in D} \mathbf{P}(A_i \cap B_j)$$

$$= \sum_{x_i \in C} \mathbf{P}(A_i) \cdot \sum_{y_j \in D} (B_j) = \mathbf{P}\{X \in C\}\mathbf{P}\{Y \in D\}.$$

373. It is not necessary. Still, the answer itself is not enough. It is easy to construct a counterexample for two discrete random variables.

374. $\mathbf{E}(S_k|\mathcal{F}_n) = S_{k \wedge n}$ a.e., where $k \wedge n = \min\{k, n\}$.

Appendix
Table of Normal Distribution Function Φ of $N(0, 1)$

t	0.00	0.01	0.02	0.03	0.04	0.05	0.06	0.07	0.08	0.09
0.0	0.5000	0.5040	0.5080	0.5120	0.5160	0.5199	0.5239	0.5279	0.5319	0.5359
0.1	0.5398	0.5438	0.5478	0.5517	0.5557	0.5596	0.5636	0.5675	0.5714	0.5753
0.2	0.5793	0.5832	0.5871	0.5910	0.5948	0.5987	0.6026	0.6064	0.6103	0.6141
0.3	0.6179	0.6217	0.6255	0.6293	0.6331	0.6368	0.6406	0.6443	0.6480	0.6517
0.4	0.6554	0.6591	0.6628	0.6664	0.6700	0.6736	0.6772	0.6808	0.6844	0.6879
0.5	0.6915	0.6950	0.6985	0.7019	0.7054	0.7088	0.7123	0.7157	0.7190	0.7224
0.6	0.7257	0.7291	0.7324	0.7357	0.7389	0.7422	0.7454	0.7486	0.7517	0.7549
0.7	0.7580	0.7611	0.7642	0.7673	0.7704	0.7734	0.7764	0.7794	0.7823	0.7852
0.8	0.7881	0.7910	0.7939	0.7967	0.7995	0.8023	051.8	0.8078	0.8106	0.8133
0.9	0.8159	0.8186	0.8212	0.8238	0.8264	0.8289	0.8315	0.8340	0.8365	0.8389
1.0	0.8413	0.8438	0.8461	0.8485	0.8508	0.8531	0.8554	0.8577	0.8599	0.8621
1.1	0.8643	0.8665	0.8686	0.8708	0.8729	0.8749	0.8770	0.8790	0.8810	0.8830
1.2	0.8849	0.8869	0.8888	0.8907	0.8925	0.8944	0.8962	0.8980	0.8997	0.9015
1.3	0.9032	0.9049	0.9066	0.9082	0.9099	0.9115	0.9131	0.9147	0.9162	0.9177
1.4	0.9192	0.9207	0.9222	0.9236	0.9251	0.9265	0.9279	0.9292	0.9306	0.9319
1.5	0.9332	0.9345	0.9357	0.9370	0.9382	0.9394	0.9406	0.9418	0.9429	0.9441
1.6	0.9452	0.9463	0.9474	0.9484	0.9495	0.9505	0.9515	0.9525	0.9535	0.9545
1.7	0.9554	0.9564	0.9573	0.9582	0.9591	0.9599	0.9608	0.9616	0.9625	0.9633
1.8	0.9641	0.9649	0.9656	0.9664	0.9671	0.9678	0.9685	0.9693	0.9699	0.9706
1.9	0.9713	0.9719	0.9726	0.9732	0.9738	0.9744	0.9750	0.9756	0.9761	0.9767
2.0	0.9772	0.9778	0.9783	0.9788	0.9793	0.9798	0.9803	0.9808	0.9812	0.9817
2.1	0.9821	0.9826	0.9830	0.9834	0.9838	0.9842	0.9846	0.9850	0.9854	0.9857
2.2	0.9861	0.9864	0.9868	0.9871	0.9875	0.9878	0.9881	0.9884	0.9887	0.9890
2.3	0.9893	0.9896	0.9898	0.9901	0.9904	0.9906	0.9909	0.9911	0.9913	0.9916
2.4	0.9918	0.9920	0.9922	0.9925	0.9927	0.9929	0.9931	0.9932	0.9934	0.9936
2.5	0.9938	0.9940	0.9941	0.9943	0.9945	0.9946	0.9948	0.9949	0.9951	0.9952
2.6	0.9953	0.9955	0.9956	0.9957	0.9959	0.9960	0.9961	0.9962	0.9963	0.9964
2.7	0.9965	0.9966	0.9967	0.9968	0.9969	0.9970	0.9971	0.9972	0.9973	0.9974
2.8	0.9974	0.9975	0.9976	0.9977	0.9977	0.9978	0.9979	0.9979	0.9980	0.9981
2.9	0.9981	0.9982	0.9982	0.9983	0.9984	0.9984	0.9985	0.9985	0.9986	0.9986
3.0	0.9987	0.9987	0.9987	0.9988	0.9988	0.9989	0.9989	0.9989	0.9990	0.9990
3.1	0.9990	0.9991	0.9991	0.9991	0.9992	0.9992	0.9992	0.9992	0.9993	0.9993
3.2	0.9993	0.9993	0.9994	0.9994	0.9994	0.9994	0.9994	0.9995	0.9995	0.9995
3.3	0.9995	0.9995	0.9995	0.9996	0.9996	0.9996	0.9996	0.9996	0.9996	0.9997
3.4	0.9997	0.9997	0.9997	0.9997	0.9997	0.9997	0.9997	0.9997	0.9997	0.9998

t	1.282	1.654	1.960	2.326	2.576	3.090	3.291	3.891	4.417
$\Phi(t)$	0.90	0.95	0.975	0.99	0.995	0.999	0.9995	0.99995	0.999995
$2(1-\Phi(t))$	0.20	0.10	0.05	0.02	0.01	0.002	0.001	0.0001	0.00001

© The Author(s), under exclusive license to Springer Nature Switzerland AG 2025
J. Misiewicz, *A One-Semester Course on Probability*, Springer Undergraduate Mathematics Series, https://doi.org/10.1007/978-3-031-86681-4

References

1. Banach, S., Kuratowski, K.: Sur une généralisation du probleme de la mesure. Fundam. Math. **14**, 127–131 (1929)
2. Banach, S., Tarski, A.: Sur la décomposition des ensembles de points en parties respectivement congruentes. Fundam. Math. **6**, 244–277 (1924)
3. Billingsley, P.: Probability and Measure, 3rd edn., Chaps. 1 & 2. Wiley, New York (1995). ISBN:0-471-00710-2
4. Borovkov, A.A.: Probability Theory. Springer, Heidelberg (2009). ISBN 978-1-4471-5200-2
5. Dudley, R.M.: Real Analysis and Probability. Wadsworth, BNrooks & Cole, Pacific Grove (1989)
6. Feller, W.: An Introduction to Probability Theory and Applications. I and II, 3rd edn. Wiley, New York (1966)
7. Inglot, T., Ledwina, T., Ławniczak, Z.: Materials for Exercises in Probability and Mathematical Statistics (in Polish). Skrypt Politechniki, Wrocławskiej (1984)
8. Jakubowski, J., Sztencel, R.: Introduction to Probability Theory (in Polish). Script, Warszawa (2000)
9. Jaynes, E.T.: The well-posed problem. Found. Phys. **3**(4), 477–493 (1973)
10. Kallenberg, O.: Foundations of Modern Probability. Probability and its Applications. Springer, New York (1997). ISBN:0-387-94957-7
11. Letac, G.: Integration and Probability Exercises and Solutions, Springer, Berlin (1995)
12. Mazurkiewicz, S., Sierpiński, W.: Sur un ensemble superposable avec chacune de ses deux parties. C. R. Acad. Sci. Paris **158**, 618–619 (1914)
13. Parthasarathy, K.R.: Probability Measures on Metric Spaces. Academic Press, New York (1967)
14. Plucińska, A., Pluciński, E.: Elements of Probabilistics (in Polish). PWN, Seria Matematyka dla Politechnik. Warszawa (1981)
15. Prokchorow, A.W., Uszakow, W.G., Uszakow, N.G.: Exercise Manual in Probability Theory: Basic Concepts, Limit Theorems, Stochastic Processes (in Russian). Nauka, Moskwa (1986)
16. Stirling, J.: Methodus Differentialis. Balliol College, Oxford, Typis Gul. Bowyer. (1730)
17. Stojanow, J., Mirazczijski, I., Ignatow, C., Tanuszew, M.: Exercise Manual in Probability Theory. Mathematics and its Applications, vol. 32. Springer, Berlin (1988). ISBN:9789027726872
18. Vitali, G.: Sul problema della misura dei gruppi di punti di una retta. Gamberini e Parmeggiani, Bologna (1905)
19. Weaver, W.: Lady Luck: The Theory of Probability. Dover, Garden City (1982). ISBN:0486243427

Index

A
Absolute moment, 104
Algebra of sets, 9
Almost sure convergence, 147
Average value, 81

B
Bayes' Formula, 31
Bernoulli distribution, 57
Bernoulli's Weak Law of Large Numbers, 152
Bernoulli trials, 34
Bertrand's paradox, 26
Beta distribution, 64
Binomial distribution, 57
Borel–Cantelli Lemma, 42
Buffon's needle, 25

C
Cantor function, 68
Cantor set, 102
Carathéodory's Theorem, 159
Cauchy distribution, 65
Central Limit Theorem, 153
Central moment, 104
Characteristic function, 121
Chebyshev's inequality, 108
Chebyshev's Weak Law of Large Numbers, 152
Chinczyn's Law of Large Numbers, 147
Classical definition of probability, 21
Combination, 4
Complete measure, 13
Complete probability space, 13

Conditional expectation, 173
Conditional probability, 28
Convergence almost everywhere, 147
Convergence in distribution, 131
Convergence in probability, 145
Convergence with probability 1, 147
Convolution of densities, 77
Convolution of distributions, 77
Copula, 113
Correlation factor, 111
Covariance, 111
Covariance matrix, 112
Crude moment, 103
Cumulative distribution function, 52

D
De Moivre–Laplace Theorem, 154
Density of distribution, 60
Density of random variable, 60
Devil's staircase, 68
Dirac delta measure, 13
Discrete joint distribution, 74
Discrete random variable, 52
Distribution function, 52
    of random vector, 75
Distribution of random variable, 50

E
Equality almost everywhere, 51
Esperance, 81
Events
    independent, 33
    pairwise independent, 33

Index

Expected value, 81, 82
  of continuous variable, 98
  of simple variable, 82
Exponential distribution, 62

F
Fatou's Lemma, 94
Field of sets, 9
Finite measure, 157
Fourier transform, 121
Function
  measurable, 47

G
Gamma distribution, 63
Gaussian distribution, 65
Geometric distribution, 58
Geometric probability, 25

H
Hölder's inequality, 106
Hypergeometric distribution, 59, 83

I
Impossible event, 13
Independent random variables, 71

J
Jensen's inequality, 106

K
Kolmogorov's First Strong Law of Large Numbers, 150
Kolmogorov's Second Strong Law of Large Numbers, 150
Kolmogorov's Zero-One Law, 143
Kurtosis, 105

L
Law of Total Probability, 29
Lebesgue Decomposition Theorem, 70
Lebesgue's Dominated Convergence Theorem, 94
Lebesgue's Monotone Convergence Theorem, 93
Lebesgue–Stieltjes Integral, 100
$\liminf_{n\to\infty} A_n$, 41

$\limsup_{n\to\infty} A_n$, 41
Lindeberg–Lévy Central Limit Theorem, 153

M
Markov's inequality, 110
Markov's Weak Law of Large Numbers, 146
Mathematical hope, 81
Mean value, 81
Measurable
  function, 47
Measure
  absolutely continuous, 172
  signed, 171
Measure extension theorem, 162
Median, 103
Moment, 103
Multidimensional random variable, 74
Multinomial distribution, 58
Multivariate Gaussian distribution, 116
Multivariate normal distribution, 116

N
Negative binomial distribution, 59
Non-measurable sets, 18
Normal distribution, 65

O
One-point distribution, 56
One-point measure, 13
Outer measure, 158

P
Pareto distribution, 105
Pascal distribution, 59
Percentiles, 103
Poisson distribution, 58
Probability, 13
  distribution, 50
Probability measure, 13

Q
Quantile, 103

R
Radon–Nikodym Theorem, 172
Random event, 13
Random variable, 45, 47
Random vector, 74

Rare events, 144
Raw moment, 103
Ring of sets, 9

S
Sample space, 12
Schwarz's inequality, 105
σ-algebra of sets, 9
σ-field of sets, 9
σ-finite measure, 170
σ-ring of sets, 9
Simple random variable, 52
Single point distribution, 56
Singular distribution, 68
Space of elementary events, 12
Standard deviation, 104
Stirling's formula, 37
Sure event, 13

T
Tail event, 144
Two-point distribution, 57
Types of random variables, 68

U
Uniform distribution, 62
Upper limit, 40

V
Variance, 104
Variation with repetition, 3
Variation without repetition, 4

W
Weak convergence, 131

The manufacturer's authorised representative in the EU is Springer Nature Customer Service Centre GmbH, Europaplatz 3, 69115 Heidelberg, Germany. If you have any concerns regarding our products, please contact ProductSafety@springernature.com

Printed and bound by CPI Group (UK) Ltd, Croydon, CR0 4YY

26/03/2026

02078952-0002